URBAN ARCHIPELAGO

A VOLUME IN THE SERIES

Environmental History of the Northeast

EDITED BY

Anthony N. Penna and Richard W. Judd

URBAN ARCHIPELAGO

An Environmental History of the Boston Harbor Islands

PAVLA ŠIMKOVÁ

University of Massachusetts Press
Amherst and Boston

Printed in the United States of America

ISBN 978-1-62534-597-4 (paper); 596-7 (hardcover)

Designed by Sally Nichols
Set in Freight and Matrix
Printed and bound by Books International, Inc.
This book is a revised version of the dissertation submitted by the author at Ludwig-Maximilians-Universität Munich in 2018.

Cover design by Rebecca Neimark, Twenty-Six Letters
Cover art by Union News Company. *A bird's-eye view of Boston Harbor,* c. 1899.
Courtesy of Geographicus Rare Antique Maps. CC 1.0

Library of Congress Cataloging-in-Publication Data

Names: Šimková, Pavla, 1986– author.
Title: Urban archipelago : an environmental history of the Boston Harbor Islands / Pavla Šimková.
Description: Amherst : University of Massachusetts Press, [2021] | Series: Environmental history of the Northeast | Includes bibliographical references and index.
Identifiers: LCCN 2021017195 (print) | LCCN 2021017196 (ebook) | ISBN 9781625345967 (hardcover) | ISBN 9781625345974 (paperback) | ISBN 9781613768723 (ebook) | ISBN 9781613768730 (ebook)
Subjects: LCSH: Archipelagoes—Massachusetts—History. | Human ecology—Massachusetts—Boston Harbor Islands—History. | Urban ecology (Sociology)—Massachusetts—Boston. | Boston Harbor Islands (Mass.)—Environmental conditions. | Boston Harbor Islands (Mass.)—History. | Boston (Mass.)—History.
Classification: LCC GF504.M4 S56 2021 (print) | LCC GF504.M4 (ebook) | DDC 304.209744/61—dc23
LC record available at https://lccn.loc.gov/2021017195
LC ebook record available at https://lccn.loc.gov/2021017196

British Library Cataloguing-in-Publication Data
A catalog record for this book is available from the British Library.

MÝM RODIČŮM

For my parents

CONTENTS

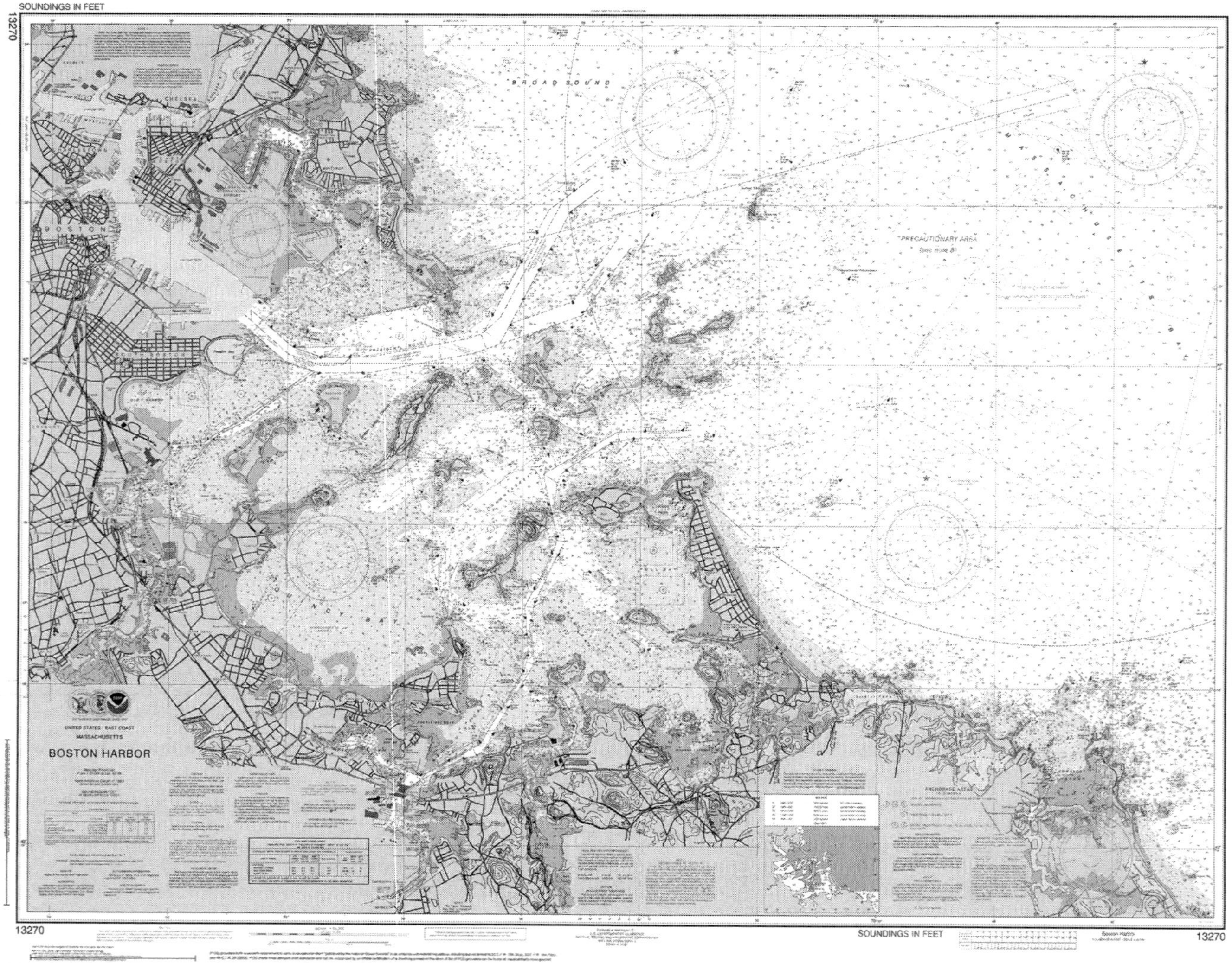

FIGURE 1. Current naval chart of Boston Harbor.—*Provided by NOAA Office of Coast Survey.*

ACKNOWLEDGMENTS

No man—and no book—is an island entire of itself. It is a pleasure to thank here the many people and institutions who have made this book possible.

This work started as a dissertation at the Rachel Carson Center for Environment and Society at the Ludwig-Maximilians-Universität (LMU) in Munich. The RCC has been at all times a stimulating intellectual environment that has facilitated encounters with scholars working on topics related to mine. My fellow doctoral students from the RCC's Environment and Society program were a wonderful community and a constant reminder that I was not alone in the ups and downs of dissertation writing. This work has, during its various stages from the first two-page draft to the book proposal, profited considerably from the discussions at the "Oberseminar" in Schönwag. Above all, I want to thank my supervisors, Christof Mauch and Uwe Lübken, who supplied invaluable feedback on my work and generously shared their knowledge and time.

Colleagues at my other institutional home, Collegium Carolinum, have always been a source of encouragement and camaraderie. Robert Luft has been a mentor and a friend for the past ten years: without his support and understanding, I would never have finished even my master's thesis, let alone a dissertation.

This project would have been impossible without the kind assistance of Boston archivists and librarians. I want to thank Marta

Crilly of the Boston City Archives, Sean Fisher of the Massachusetts Department of Conservation and Recreation, Karen Graham of the Massachusetts Water Resources Authority Library, John Hannigan of the Massachusetts Archives, Jessica Holden of the University of Massachusetts Archives and Special Collections, Silvia Mejia of the Massachusetts State Library, Jane Parr and Sarah Pratt of the Howard Gotlieb Archival Research Center at Boston University Libraries, Laura Peimer of the Schlesinger Library at Harvard University, and Katie Zimmerman of the Massachusetts Institute of Technology (MIT) Libraries. The Norman B. Leventhal Map and Education Center at the Boston Public Library was an indispensable source of historical maps of Boston Harbor, and the staff of the Boston Athenaeum helped me locate several rare nineteenth-century Boston Harbor guides. My trips to Boston in 2016 and 2017 were made possible by a scholarship from LMU's Graduate School of Language and Literature.

During my research in Boston, I benefited from the generous help of a number of people. Maria Judge welcomed me in her lovely home in Medford every time I came to Boston and helped me make first contacts in the harbor islands community. Kathy Abbott, Al Hebb, Beth Jackendoff, George Marsh, Kate Rivera, Larry Rosenblum, Stephanie Schorow, and Jan Wampler shared with me important insights about the past and present of the Boston Harbor Islands. I owe a special debt of gratitude to Suzanne Gall Marsh and the Friends of the Boston Harbor Islands. Suzanne has been a tireless advocate for the harbor islands for more than forty years, and both this book and its author have benefited immensely from her vast knowledge about the harbor islands, her generosity, and her contagious enthusiasm.

Conversations with fellow scholars, colleagues, and friends have been vital for the development of this book. My thanks go to Lenka Holcnerová, Karsten Holste, Dietlind Hüchtker, Travis Klingberg, Matthew Klingle, Patrick Kupper, Katherine Morrissey, Robert Neustädter, Chris Pastore, Jana Piňosová, Katie Ritson, Helen Rozwadowski, Arnošt Štanzel, and Anna Wöbse, who have read parts of the manuscript, provided helpful feedback, or, indeed, had the patience to listen to my constant islands babble. Jim O'Connell and an anonymous reviewer engaged with the manuscript in an extremely attentive and insightful way and made suggestions that have improved the text substantially. Any remaining flaws are mine alone.

I am grateful to Richard Judd and Anthony Penna for accepting my book into their series. Brian Halley remained professional even when faced with my most wide-eyed questions and guided the manuscript expertly through the review and publication process. Thanks are due to his colleagues at the University of Massachusetts Press Rachael DeShano and Sally Nichols, for preparing the manuscript for publication, and to Annette Wenda, for her careful copyediting.

My greatest thanks are to my parents, Rafaela Šimková and Zdenek Šimek. Without their unwavering and unconditional support, this book could never have been written.

URBAN ARCHIPELAGO

INTRODUCTION

"Minutes away, worlds apart" are the Boston Harbor Islands from Boston, according to the motto of the national and state park that encompasses all thirty-four of them.[1] Some of the islands and peninsulas are ragged rocks teeming with typical Atlantic Coast wildlife such as American oystercatchers or harbor seals. Others have the appearance of wooded hills rising gently out of the water or of elaborate Frederick Law Olmsted–designed parks. Together, the islands form a surprisingly diverse ensemble of natural-looking landscapes just a stone's throw away from one of the nation's major urban centers. It takes the ferryboat less than one hour to get to even the remotest of them; most lie within a ten-mile radius from the city. While several decades ago even most Bostonians had only a vague idea that there were any islands out there in the notoriously polluted Boston Harbor, today the picture is a very different one. The 1996 establishment of the Boston Harbor Islands National Recreation Area has framed the islands as an attractive part of Boston's open space. The islands have their own information pavilion on the immensely popular Rose Kennedy Greenway in the middle of the waterfront district. Come summer season, all Boston-area magazines and newspapers run articles about the pleasures to be had on the islands, and every recent tourist guide lists them as one of Boston's top-ten sights, alongside Fenway Park and the Freedom Trail.[2]

When you succumb to the urge to complete your list of Boston's must-sees and board one of the hourly morning ferries leaving from Long Wharf to the harbor islands, chances are you will—after a twenty-minute ride down the harbor past waterfront condominiums and art galleries on the right and the low-lying banks of Logan International Airport on the left—find yourself on Spectacle Island. Spectacle is a 105-acre island sitting at the entrance to Boston's inner harbor, approximately four miles from the city's downtown as the boat sails and one mile from the nearest shore, the peninsula of South Boston. It is green, with gently sloping shores and two rolling hills overgrown with native shrubs and trees. Except for the pier, a small visitor center, and three wooden lookout pavilions, nothing seems to suggest that this natural place has ever known disruptive human presence. Still, it is one of the "hubs" of the national and state park: an island with frequent boat service and large visitation numbers that functions as a major tourist destination and as a starting point for further island hopping. Once there, a range of island activities presents itself. You can stroll along one of the gravel paths hugging the island and enjoy the views of the harbor and of Boston's skyline in the distance. You can sit on the porch of the visitor center and listen to a live jazz concert, every Sunday during the summer season. If you trust the water-quality reports that assure you that it is indeed perfectly safe to bathe in the harbor, you might even take a swim at the island's sandy beach. Or, if you are especially hungry for knowledge, you can take a ranger-led tour of the island and find out about its rich and—given its present impeccably natural appearance—surprising history. The idyllic park, you learn, sits atop a seventy-foot layer of refuse that had been ferried over from Boston in the first half of the twentieth century. "We're still a dump," says Ranger Kathie of the Department of Conservation and Recreation, which manages the island jointly with the National Park Service, during my visit, brandishing a bleached horse's jawbone salvaged from the landfill. Over the course of more than fifty years, Spectacle daily received hundreds of tons of rags, bones, broken china, and rotting vegetables collected in the streets of Boston. Some of it was processed and then burned; some of it was dumped on the island as it was. Only in the 1990s and early 2000s was the former garbage dump covered with 3.5 million cubic yards of fill, landscaped, planted, and opened as a park.

FIGURE 2. Spectacle Island in July 2017, looking south toward Moon Island. Today, nothing in the island's natural appearance suggests to visitors its varied history. —*Photo by the author.*

Today, nothing in Spectacle Island's parklike appearance suggests this colorful past. Just like its original shape had been obliterated by a thick sheet of trash, so has the park completely obscured the seeping landfill beneath it. Yet there is one thing the trash heap of the past and the park of the present have in common: none of them would have existed if it were not for the city of Boston.

Shaping Spectacle, Shaping Boston

If it were not for Boston, Spectacle Island would not look the way it does today. In 1630 when the English Puritans of John Winthrop's party sailed into the calm bay that was about to become known as Boston Harbor, Spectacle was a conspicuously shaped forty-two-acre island, its two drumlins connected by a narrow spit that made it look like a pair of eyeglasses. Like the other islands in the early days of the colony, Spectacle was soon appropriated by the people of Boston as a source of wood and a safe cattle pasture. In the first half of the nineteenth century, it experienced brief fame as a popular spot for Sunday outings and even had two inns built on it. Starting in the 1850s, the island hosted a succession of waste-processing businesses, the first of them a factory rendering dead horses and cows. From that time on, Spectacle became the most extreme but still a fairly typical manifestation of the "dumping ground" use to which Boston put many of its harbor islands. The horse-rendering plant was followed by a glue factory, a garbage-reduction plant, and finally the landfill that buried the northern portion of the island under seventy feet of trash. When the landfill closed in the 1950s, the island fell into disuse, an embodiment of the general lack of interest that characterized Boston's attitude toward the islands and the harbor during this period. In the 1980s and 1990s, popular interest in the harbor and the islands increased again, this time because of the recreational opportunities they presented. In the wake of this wave of attention, Spectacle Island emerged as a symbol of the renewal and restoration of the harbor. As it prepared for its new role as an island park, Spectacle swelled to more than twice its original size, lost its characteristic shape, and was built up to become the highest point in the harbor. All of these changes were attributable to the transforming influence of Boston. The town appropriated the island early on and over the next four hundred years put it to various uses that reflected its current needs—and in the process altered the small piece of land in the harbor quite beyond recognition.

This relationship, although admittedly asymmetric, has not been a one-way street: if it were not for Spectacle Island, Boston would not look the way it does today, either. Those 3.5 million cubic yards of earth that were dumped on the island in order to transform it from an eroding landfill into an orderly park had been dug out from underneath Boston's waterfront and from the harbor floor. In the 1990s, Boston launched the mammoth construction project of building a third harbor tunnel and depressing a highway that until then cut the city's downtown in half. The project, colloquially known as the Big Dig, has radically reshaped Boston's waterfront district, making it into a pedestrian-friendly, livable part of town, and would have been extremely difficult to carry out without the option of depositing the mountain of excavated clay and sediment on Spectacle Island. As Boston transformed the island, so did, in a way, Spectacle Island transform the city.

For many of the tourists who today arrive in Boston, Spectacle Island has advanced to a representative of all the Boston Harbor Islands. This impression is, from a historical point of view, largely accurate: Spectacle may be an extreme example of the changes the islands experienced over the years, but most of them went through similar transformations. Several of the islands were swallowed whole by the city: Governors, Bird, and Apple Islands have all disappeared under the runways and terminals of the growing Logan Airport. Others, like Deer, Castle, Moon, and Nut Islands, have become appendices of the mainland, attached to it by fill or causeways. Georges Island has practically become one with a large fort that takes up most of its area. Some of the islands, used as quarries for gravel, have disappeared entirely, like the former island of Nixes Mate, which is today only a shoal marked by a wooden beacon. Others have been reduced to a fraction of their former size, like Sheep Island in Hingham Bay. There is not one island in the harbor that has not been shaped in one way or another by the changing demands and practices of Boston's residents. Spectacle Island is in this sense but a prominent example of a general phenomenon. Its history captures the arc of development of the Boston Harbor Islands as a whole. This is why this book uses it at times as a stand-in for all the harbor islands.

The Islands and Their City

The dramatic way Boston has transformed its natural environment has captivated historians for decades. This is hardly surprising: although a number of American cities embarked in the nineteenth century on a large-scale rearrangement of their natural settings, the radical way Boston has coped with its original confined location on a peninsula scarcely one square mile in size still commands attention.[3] More than a half of the city we know today stands on made land; in some areas this ratio is as high as three to one.[4] In the course of its history, Boston has cut down its hills to fill its tidal flats, has been constantly changing the shape of its shoreline and the depth of its harbor by dredging and filling, and has completely altered the nature of its rivers. In the nineteenth century, it built the fashionable residential district of Back Bay over a vast expanse of mudflats and stagnant pools of water, and more recently it paved over 1,413 acres of East Boston Flats to build and expand the airport. In the late twentieth century, it was the site of two of the largest public projects in the United States' history, the aforementioned Big Dig and the $3.9 billion Boston Harbor cleanup. The city's changing topography has become the subject of a number of historical and geographical studies.[5] Yet curiously, Spectacle and the other thirty-three Boston Harbor Islands are the one part of the city's environment that hardly ever receives any attention at all. Recently, historians have started to emphasize the connection between Boston and its harbor, the source of the town's wealth and the link connecting it to the Atlantic world, but the islands remain largely invisible.[6] If any mention is made of them at all, either they tend to be treated summarily as marginal, isolated places with little or no attachment to the city, or else they are mentioned one at a time as specific sites defined only by their current use while their relationship to the city and to one another remains obscure.[7] Yet there are parallels that can be drawn between all the islands: regardless of their various shapes, sizes, and distances from the city, Boston has understood and treated them as one ensemble throughout its history. A closer look at the islands' historical relationship with Boston reveals them as places that are essential for understanding Boston's relationship with its environment and the way this relationship has changed over time. This book argues that without considering its harbor islands, we cannot fully understand the environmental history of Boston.

The Boston Harbor Islands are unique when regarded only in the Boston context. A widening of the perspective, however, reveals dozens of similar islands lining the coasts of North America and, indeed, the world. These are islands close to coastal urban centers, islands that have been drawn into "their" cities' urban networks, went through similar developments, and were subjected to similar influences, all the while being perceived as isolated, different, other. They form their own distinct category of places: they can most accurately be called urban islands.[8]

Urban islands have historically been special places. They are close enough to the mainland so as to be drawn into urban infrastructure, but the fact that they are surrounded by water has in the past three centuries given them the aspect of isolation and remoteness. For a place to be seen as an island carries with it rich cultural implications, most important associations of mystery, romance, and otherworldliness. These attributions have often enough been applied to even so near and familiar places as urban islands, simply by virtue of their insularity. On a more material level, urban islands also seem to have attracted certain very specific uses over the years, uses that profited from and exploited their island locations. This book asks what it means for a place to be an urban island, what this specific geographic situation has meant for the way the islands have historically been perceived and used, and how these perceptions and uses have changed over time.

The history of the changing position of islands within urban networks is by its nature a place-based story. It is therefore important to tell it focusing on particular islands near a particular city. Only in this way can the complex terrain of the islands' many roles for "their" city be mapped in all its details and nuances. In few places are the diverse uses that urban islands have been put to as manifest as in Boston. Due to the city's—by North American standards—relatively long history, its treatment of its islands is a representative sample of most of the ways other North American coastal cities have used their islands over time. The Boston Harbor Islands' historical incarnations as quarries, pastures, quarantines, landfills, fortresses, and parks have been repeated all along the urban coasts of North America. Recounting the history of the Boston Harbor Islands thus offers a look into the past of America's urban islands in a nutshell.

Urban Archipelago is a book about the relationship between Boston and the Boston Harbor Islands: about the ways the city has perceived and used the islands since its founding until the present day and about the ways both the islands and Boston have changed in the process. Over four hundred years, private citizens, town councilors, city and state politicians, public agencies, the maritime community, landscape architects, businessmen, city planners, Boston Harbor boosters, ecologists, and the residents of the metropolitan area have had various ideas, often conflicting ones, about what the islands of Boston Harbor were good for and how they should be used and by whom, and they have tried to impose their vision on the physical environment of the islands. All of these visions, however, reflected at the same time their proponents' ideas about the city of Boston itself, what it is and what it should be.

The Other Islands

Over much of Boston's history, the city's principal perception of the harbor islands was that of otherness. While the colonial town thought them as much a part of its space as the Boston peninsula itself and the water in between rather a connection than a separation, later times regarded them as a polar opposite of the city, a container for what it discarded and a magic box providing what it lacked. They were the trash heap of the sanitary city of the nineteenth century, the redemption of the ailing urban core in the era of the 1960s urban renewal, and the open space of the sprawling modern metropolis craving recreation. Yet no matter how different they were from the city's congested streets in their physical form and no matter how isolated they seemed by the waters of the harbor, the Boston Harbor Islands have at any given time been mirroring the city's development and have been firmly embedded in its urban fabric. Just like the surrounding fields and woods have been appropriated by the expanding city and made into suburbs, industrial sites, and parks, so, too, have the islands become a part of the city—but in a unique way. They have been incorporated into the city's body through their use while at the same time being perceived and presented as not belonging to it.[9]

This dual relationship of perceived otherness and functional attachment has had very real consequences for the islands' environment. Their perceived separation from Boston made them into ideal places for uses either undesirable or unattainable in the city itself, Boston's veritable "other." Yet at the same time, whatever Boston's current needs and values have been, the harbor islands would be developed accordingly, whether it was as pastures, refuse dumps, or urban parks. Even when the islands' appearance suggested a complete separation from the urban space, the original impulse for the changes they went through could almost always be traced back to Boston. In this respect, there has never been much difference between a marina or a sewage tank, a lighthouse or a quarantine hospital. They have all been built for the benefit of the city, to accommodate its current needs.

To argue, however, that the Boston Harbor Islands have been only reflections of the metropolis across the water would be denying that they are places in their own right. Spectacle, like many of the other harbor islands, was from the colonial times until the 1950s home to a small community of people, farmers at first and later employees of the many waste businesses that settled on the island. The islanders, numbering at times as many as one hundred people, felt an attachment to their island that was largely incomprehensible to the residents of the mainland. Where mainlanders saw a horrible, foul-smelling wasteland, Spectacle's inhabitants perceived a pleasant, verdant island with fresh sea breezes. It was a place where you went swimming and boating with your neighbors in the summer and skating on a small pond in the winter; it was a place where you went to work in the waste-processing factory and grew vegetables in your patch of garden; it was home.[10] The values and perceptions of the islanders often ran contrary to the city's interests, and sometimes they even made a difference for the islands' development. More often than not, however, the opinions and values coming from Boston prevailed over those of the island populations, and the islanders' affection for their harbor homes played little role in the islands' fortunes.

The close relationship between islands and adjacent urban settlements on the mainland does not apply to all coastal islands. Some, such as Long Island, Aquidneck Island in Narragansett Bay, or Martha's Vineyard, are so large as to function like independent units in their own right or are so accessible from the mainland as to constitute an indiscernible part of it. On the other end of the spectrum, weather-beaten rocks jutting out of the water have rarely been used for purposes other than the occasional lighthouse, however close to mainland urban centers they may have been. Some coastal islands are too far from the mainland to function as its extension: this is the case with California's Channel Islands. There are scores of islands and seacoasts that were shaped by human activity in fundamental ways without being drawn into the orbit of a large population center on the mainland. At times, they share vital parts of the "working landscapes turned recreational areas" story of the Boston Harbor Islands. Prominent and well-researched examples include California's Point Reyes National Seashore north of San Francisco or the Apostle Islands in the southwest corner of Lake Superior.[11] The focus of this book, however, is on the specific interactions between coastal islands and urban centers that treat "their" islands with characteristic ambivalence: professing their isolation and otherness just as they draw them ever deeper into the urban networks.

Islandness, Remoteness, and Lines in the Sand

There are several themes running through the shared history of Boston and its urban islands that inform the basic dichotomy of the islands' simultaneous separation from and attachment to the city. One of them is the very notion of "islandness": the fact that islands as places and landscapes are perceived and treated differently from other sites precisely because they are islands. "The truth is, if you came upon the 50 acres of Grape Island in some suburb, you would be underwhelmed. It would seem like just a small stretch of charming, hilly farmland," *New York Times* reporter Carey Goldberg remarked in her article about the Boston Harbor Islands in 2000.[12] The idea that places that lie beyond the edges of continents are special and apart but still somehow reflect and complement the mainland to which they do not belong has wound its way through much of Western civilization's thinking about islands.[13] In Boston the perception that island settings have an intrinsic value that landlocked spots lack gained currency in the nineteenth century with its appreciation of healthful sea breezes and rose to further prominence as the islands gradually ceased to be places of work and were reframed as places of leisure in the second half of the twentieth century. Geologically speaking, the Boston Harbor Islands are a half-submerged field of glacial drumlins formed during the last ice age. This is not a unique quality: the entire landscape of Boston Basin has been shaped by advancing and receding glaciers, and the metropolitan region is strewn with drumlins of all shapes and sizes. The fewest of them have ever been considered anything more than pleasant landscape features or sources of fill material—except for the Boston Harbor Islands, which were seen as valuable enough to be declared a national park. Neither was Spectacle Island the only landfill in Boston that has received its share of fill from the Big Dig project. Millennium Park in the neighborhood of West Roxbury sits atop the former Gardner Street landfill, which was covered up and converted into a park at about the same time as Spectacle. Although roughly of the same size and almost identical in appearance to Spectacle, Millennium Park draws no crowds. It is known to a handful of locals, and it is hard to imagine that anyone would pay twenty-five dollars to go and see it—as hundreds of people do every summer weekend to get to Spectacle. The question of how the simple fact that a place is surrounded by water instead of houses or fields changes the way it is regarded and used joins with the notion of otherness of the islands to form one of the key themes of this book.

Islandness sometimes means not only a perceived difference, though, but a real one. For people without a boat, islands are physically isolated places that are difficult to access. This quality makes them seem remote and detached regardless of whether they are that in absolute terms or not. The topic of remoteness is one that surfaces throughout the whole history of the islands. Actual distances do not play much of a role here: even those islands that are farthest from Boston are about the same distance from the statehouse on Beacon Hill as the already mentioned outlying neighborhood of West Roxbury and fairly close to the city in absolute terms. Perceived remoteness, on the other hand, is quite a different thing: the islands have been drawn closer and pushed farther from the city depending on their current use and perceived desirability. Today, the islands are more present in the city than ever before, or rather in a different way: whereas a hundred years ago the main sign of the islands' presence was the odors wafting from Spectacle Island's waste businesses to South Boston, today the islands are advertised as a tourist destination in Boston's busy waterfront district. The fluidity of the islands' distance from Boston and the question of access are important factors in deciphering the islands' relationship to Boston.

The history of the Boston Harbor Islands is in a sense a history of zoning, of drawing lines, and of the challenges of trying to create fixed boundaries in the ever-changing littoral environment. One aspect of Boston's historical interaction with its islands and its shore is the centuries-long and often futile struggle to pin them down, to confine them to one place and one shape. Just like the shoreline is merely a mental construction in the fluid landscape governed by ebb and flow, the amorphous environment of tidelands, forming shoals, and vanishing spits, erosion, sedimentation, and tidal currents, so have the shapes of the islands changed over time, often as a result of human activity. The attempts at describing and controlling the essentially uncontrollable inform the discussion of Boston's relationship to its harbor and islands in this book.

The Boston Harbor Islands lie on the threshold between land and ocean, between urban and natural space. Similarly, this book is situated at the intersection of disciplines: it is mostly informed by urban environmental history while also building on the growing literature in the environmental history of coastal regions.[14] Environmental histories of coastal cities, although they rarely accentuate islands, offer a valuable look at the ways cities on the shore have treated their urban space and their hinterlands over time.[15] From historical geography, this book borrows an inquiry into the historical relationship between islands and continents.[16] This book also owes much to the historical study of islands, as represented mainly by the work of John Gillis.[17] The discipline of island studies offers a thorough interrogation of the concept of islandness, both in spatial and in cultural terms. However, between urban history and island studies, small islands located near coastal cities often fall through the cracks: not even studies concerned with coastal regions seem to care much about what lies beyond the water's edge. Urban histories tend not to regard such islands as parts of cities, whereas island studies, which focus predominantly on larger islands and issues of resilience, risk, or climate change, often see them as too small and insignificant.[18] They may be that when considered on their own: for "their" cities, however, they represent an often invaluable resource and an instrumental part of urban networks. The example of Spectacle Island and the Boston Harbor Islands allows me to draw comparisons with other harbor cities and their islands in North America and to bring these inconspicuous yet essential parts of urban topographies back into the picture of urban environmental history.

From Puritans to Parks; or, The Four Centuries of Boston's Urban Islands

The five chapters of this book follow a story arc that spans the four hundred years of the Boston Harbor Islands and Boston's shared history and covers the islands' gradual transition from working landscapes into places of leisure.[19] It needs to be stressed that this development has been neither linear nor absolute, and the perceptions of the islands have hardly been monolithic. At any given time, conflicting views have existed of the islands and of the way they should be used: in the nineteenth century, at a time when the industrial uses of the islands reached a high point, the harbor space also emerged as a popular recreational area. Similarly today, some of the islands are still used for the older "dumping ground" purposes while simultaneously being part of the national park. Nevertheless, it is possible to detect a gradual movement from places of labor toward places of leisure, from work toward play, from an active everyday engagement to infrequent seasonal leisure activities and passive contemplation.

Despite following this transformation, this book does not want to present a narrative of a flawed and misled past leading up to a glorious present. Each of the historical uses of the islands followed the logic of their own time. In the context of their period, even those proposed and actual uses that to us, who possess the unfair advantage of knowing the outcome, seem downright absurd were in most cases guided by an honest effort to solve the problems of the day and by a genuine belief that this was the best way to do so. The creation of the present-day national park, therefore, is not treated here as a triumphant end of history, but as merely another in a succession of uses, guided almost exclusively by the perceptions and needs of the residents of the metropolis on the mainland.

The first chapter tells the story of the islands as embodied parts of the colonial town of Boston. Easily reached and located along the early town's only highway, the harbor waters, the islands formed Boston's natural hinterlands. They were at first used as sources of wood and fodder and as sheltered pastures for the colonists' cattle. Later, as the town turned its gaze toward the sea and maritime trade, they were treated as resources of ballast stones and ship masts. Connected to Boston by the water medium, they were regarded and used as the town's oceanic extensions.

The second chapter focuses on a time when the islands, although closely functionally tied to the city, became the ultimate urban fringe: places that came to house businesses, institutions, and people the city wanted out of the way. Refuse, dead animals, the sick, the poor, the criminal were all ferried out onto the harbor archipelago. These uses capitalized on the islands' perceived remoteness and isolation from the mainland. Despite these perceptions, the islands became during this time an integral part of Boston in their function, reflecting the city's development and its needs, while at the same time presenting its dark, unsavory side.

The third chapter encounters the islands as places in transition: with the decline of Boston's port and with a general lack of interest in the harbor, the islands seemed all but forgotten in the 1950s. The old remoteness-seeking facilities mostly fell into disuse, and harbor recreation stagnated. In the 1960s, however, multiple plans emerged that were supposed not only to put the islands and the harbor to new uses but first and foremost to help solve Boston's urban problems. Alarmed by a dwindling tax base, white flight to the suburbs, and a deteriorating downtown, Boston planners looked for ways of renewing the city and attracting well-to-do residents and businesses. The islands emerged at this time as seemingly vacant space, as lands of opportunity that could help redeem the old city. Contemplated, on the one hand, as sites of a new airport and even a utopian new community for fifty thousand souls and, on the other, as a haven of conservation and recreation, the islands became projection screens of the city's plans for its future.

The 1960s and 1970s were also a time when the idea of the Boston Harbor Islands as romantic, out-of-the-ordinary places that deserve protection and should be "enjoyed" rather than worked by the people of the metropolitan region caught on. The fourth chapter emphasizes the role of Edward Rowe Snow, author of popular books about Boston Harbor, in popularizing a narrative of the Boston Harbor Islands as intriguing places of mystery and adventure. His narrative cast even the heavily urbanized islands of the inner harbor in the role of isolated insular gems, and it was instrumental in separating the islands from the city in the public mind.

The fifth chapter and the epilogue trace the transformation of the harbor islands in both physical form and perception into places of leisure, a tourism asset, and a space supposedly "worlds apart" from Boston. A new environmental sensibility, simultaneous developments like the Boston Harbor cleanup, and an emerging narrative of the islands as Boston's urban wilderness helped bring about a turnaround in the way the harbor islands were perceived and used. In contrast to previous times, Boston's residents now started to relate to the islands almost exclusively through leisure. Recreation emerged as the principal way in which Bostonians interacted with their harbor islands. Yet although the islands' shape changed dramatically during this period, their underlying relationship to the city remained the same. The islands entered the twenty-first century in their four-hundred-year-old twofold role as both Boston's antidote and its mirror.

This book is centered around the claim that for much of their shared history, Boston and its urban archipelago have formed one large whole. At any given time, the islands have been shaped primarily by Bostonians' perceptions and needs. What has been changing, however, is the city itself and the manifestations of this principle. The story of Spectacle Island in particular and the Boston Harbor Islands in general allows me to explore how the ways the city understood itself have changed over time, how the changing narratives and the actual development of the islands signified the changing understanding the city had of itself, how conflicting interests have manifested themselves in conflicting narratives about the islands, and, not least, how the islands in turn have helped shape the city into its current form by allowing it more space in its own boundaries, shaped its self-image, and contributed to the solution of urban ills. The Boston Harbor Islands may be minutes away from Boston. But they have never been worlds apart.

A Note on Terminology

What I refer to as Boston Harbor throughout this book is the space in its most commonly used boundaries: the water expanse, the coast, the harbor floor, the islands, and the tidal flats landward of the imaginary line between the tip of Deer Island and Point Allerton, as well as the outer islands—the Brewsters, the Calf and Green Islands, and the Graves—and the water surrounding them. Technically, also the tidal estuaries of the rivers flowing into the harbor should be considered part of the harbor; however, they play little role in the story I tell here.

Boston was incorporated as a city in 1822. Whenever possible, I refer to the Boston before that date as *town* and after 1822 as *city*. Sometimes, however, especially when speaking about the whole span of Boston's history, it is impossible to make this distinction. In these cases I mostly opt for the term *city*. Although it technically does not apply to the pre-1822 town, it would be still more inaccurate, not to mention absurd, to refer to present-day Boston as a town.

Historically, most of the islands experienced considerable fluctuations in their name forms: for instance, what is today called Peddocks Island was at various points in time known as Pettick's, Puttock, or Peddock's Island. For the sake of clarity, I use the current name of each island as used by the agencies managing the Boston Harbor Islands National and State Park, even when speaking about historical periods when the islands may have been known under different names.

1

THE COMMONS OF BOSTON HARBOR

The Harbor Islands and Colonial Boston

Richard Mather was not only a godly man but, to his own mind, also one favored by God's Providence. After an arduous twelve-week voyage across the North Atlantic, in August 1635 the noted minister and his fellow English colonists aboard the ship *James* finally glimpsed the coast of New England, the destination of their journey. They made landfall at Monhegan Island off the coast of Maine, and after another brief respite at the Isles of Shoals, a group of rocky islands six miles offshore facing what is today the border between Maine and New Hampshire, continued on the final leg of their voyage to the new colonial town of Boston. The *James* was about to set sail toward Cape Ann when it got caught in "a most terrible storm of rain and easterly wind" that became known to future generations as the Great Colonial Hurricane. The desperate attempts of the ship's crew to stay moored and wait out the storm were in vain: the *James* lost all three of its anchors, and the gale swept the helpless ship toward the rocky coast, its sails ripped to pieces and the passengers trembling for their lives. Yet miraculously, when the storm subsided, the *James* was still afloat and everyone on board unharmed. Slowly, the tattered ship made its way southwest along the coast and in the evening came to rest in the safe embrace of Boston Harbor. As Richard Mather recalled in his journal, the ship arrived "in a most pleasant harbour, like to which I had never seen, amongst a great many of islands on every side." After the ordeal of the open sea, there could scarcely be a more paradise-like place on earth to the survivors than these sheltered waters. The *James* reached Nantasket, just inside the harbor entrance, and that same night, after the evening service, it followed the incoming tide into the deepwater channel that snaked its way among the many islands of the bay and finally arrived in Boston.[1]

Richard Mather went on to serve as the minister of Dorchester, another township on the shores of Boston Harbor, for the next thirty-three years and became the patriarch of a prominent family of Boston divines that included his son Increase Mather and his grandson Cotton Mather. Despite his exceptional position, his account of his eventful voyage to the New World shared many features with other examples of its genre. The wearisome waiting for westerly wind, the long and strenuous passage across the vast aquatic plain that occasionally transformed into a frightening landscape of welling mountains and valleys of water, and the titillating expectation of a new life in the unknown land beyond the sea were features common to many chronicles of the Atlantic passage.[2] As in Mather's tale of ordeal and deliverance, the coastal islands of New England played a prominent role in other accounts, even those written by colonists who arrived under less dramatic circumstances. Amazed by water-spouting whales, porpoises, and "grampuses as big as an ox" leaping around their ships, but equally wearied after multiple weeks of cramped quarters, inadequate food, and seasickness, the Puritans were relieved to set eyes at last on firm ground that promised solid rock beneath their feet as well as freshwater and grass for their cattle.[3] The small, rocky islands lining the New England coast were for the newcomers from the East the first sites of contact with the new continent. Places like the Georges Islands before the coast of Maine, the outlying Monhegan Island twelve nautical miles offshore, the Isles of Shoals opposite the mouth of the Piscataqua River, Thacher Island off Cape Ann, or Bakers Island at the entrance of Salem Harbor have all advanced to local landmarks well known to sailors by the early 1630s. The larger islands were also, naturally, among the first places to be settled by the English, originally as fishing outposts. When the party on the *James* that included Richard Mather approached Richmond Island off Cape

Elizabeth in 1635, they encountered there a colony of some forty fishermen, who "were sore afraid of us, doubting lest we had been French, come to pillage the island."[4] Fishing colonies were common on the Northwest Atlantic coastal islands, but those who came on the Puritan ships were looking for something other than abundant fisheries. For them, the islands were the first emissaries of the land that lay beyond: the first tokens of the fertility of the settlers' new home. Of more immediate importance, the islands helped create stretches of comparably calm coastal waters and sheltered harbors that provided refuge from the perils of the open ocean; they also signaled the welcome proximity of the continent. Thus, the sight of the Boston Harbor Islands, noted with such delight by Richard Mather, promised both relative safety and a speedy end to the rather unpleasant adventure of the Atlantic crossing.

The "most pleasant harbour" where the *James* found a safe haven had only recently become the center of the New England colonies. Five years before the *James*'s dramatic journey, in April 1630, four other English ships set sail from the Isle of Wight towards the New World. They were the *Arbella*, the *Talbot*, the *Ambrose*, and the *Jewel*, and they carried a company of several hundred settlers that included John Winthrop, the future governor of the colony of Massachusetts Bay. After a nine-week voyage, they made landfall at Salem, the principal port in the earliest days of the New England colonies. The new "planters," however, did not stay there long, for, as Winthrop's deputy Thomas Dudley noted, "Salem, where we landed, pleased us not."[5] Probably because of an outbreak of disease that plagued Salem at the time, only a few days after arrival the newcomers sent a small party to survey the coast and identify a place better suited for the new colony. As it happened, such a place lay less than twenty miles to the south: the so-called still bay within the Massachusetts Bay, which in a few years would become known as Boston Harbor.[6] The bay was sheltered, with several tributaries that provided access farther inland, and its shores seemed to have good pasturage and freshwater sources. After the initial foray, the rest of the party soon followed and in late June 1630 arrived at the shores of what was shortly to become the town of Boston.[7]

A World of Islands

When they entered the harbor, the colonists encountered a world of islands. The bay encompassed some thirty of them, varying in shape, size, and height, some overgrown with trees, others little more than sand and gravel shoals. A few of them quickly became familiar landmarks, guiding ships to the new town. When the *Arbella* and its companions cleared the dangerous rocks and rocky islands at the harbor entrance, which later came to be called the Brewsters and the Graves, they sailed through a deepwater channel past the tip of Deer Island, the northern headland protecting the harbor from the

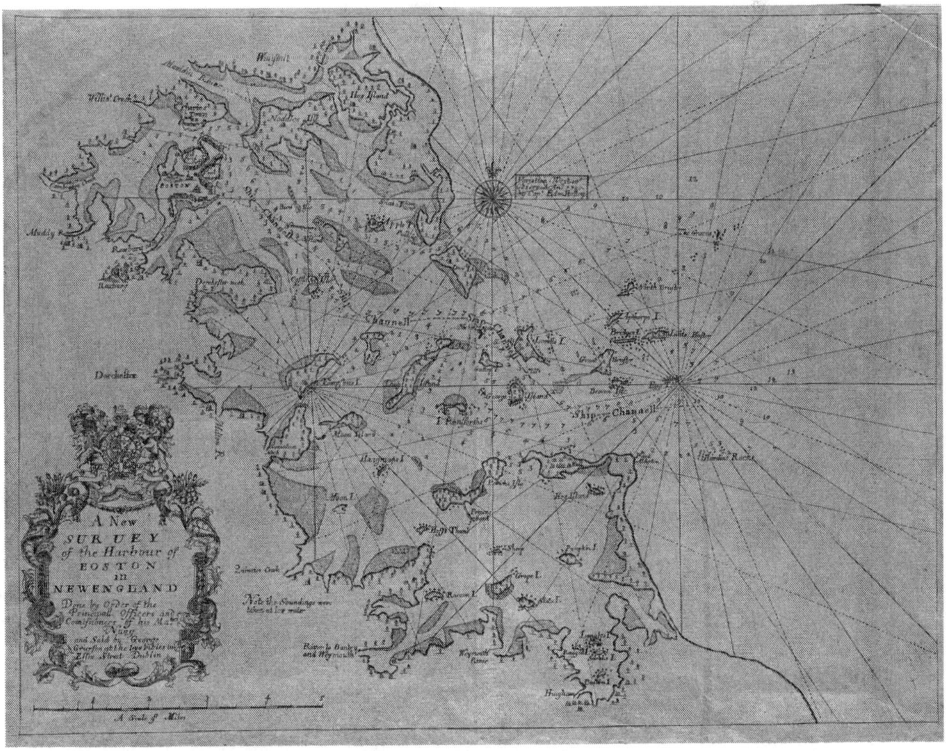

FIGURE 3. A 1708 chart of Boston Harbor from *The English Pilot*. Special attention has been given both to shipping channels and to shoals, mudflats, and other hazards. This was the coastal environment the Puritans encountered when they arrived in the 1630s.—*Map reproduction courtesy of the Norman B. Leventhal Map and Education Center at the Boston Public Library. This work is licensed for use under Creative Commons license CC BY-NC-SA 3.0.*

rough seas of the outer bay.[8] The southern headland, Point Allerton, would have been obscured by Lovells, Gallops, and Georges Islands, a group of midsize isles huddled closely together at the harbor's doorstep. After passing on the port side the small, barren island of Nixes Mate, the tip of the large and wooded Long Island, and the northern cliff of Spectacle Island, rising abruptly above the waves, and after successfully avoiding the Lower Middle, a dangerous shoal on the starboard side, the Puritan ships turned northwest and found themselves in the inner bay, lined on both sides with vast expanses of tidal flats and leading up to the destination of their journey, a peninsula at the head of the bay they called Charlestown.

The colonists of John Winthrop's party were not the first English who settled down in the bay. The islands and peninsulas of the still bay of Massachusetts were home to a handful of English immigrants who had arrived at various points in the 1620s. There was the family of David Thompson on one of the larger islands close to shore that was later named Thompson Island in his honor; there was the rather reclusive Reverend William Blackstone on the hilly Shawmut peninsula, across the water from Charlestown; and there was Samuel Maverick on Noddles Island, a man noted for his hospitality and general agreeableness that to some extent made up for his being "an enemy to the Reformation in hand."[9] Charlestown did not stay the center of the new plantation for long. Plagued by disease and dismayed by lack of drinkable water, most of Winthrop's colonists soon made good on William Blackstone's invitation and relocated to the Shawmut peninsula. The settlement they built there they called Boston. Other settlers dispersed around the bay: some went up the Mystic River and settled at Winnisimmet (today's Chelsea) and Medford. Others sailed, or rather rowed, up the Charles River and founded Watertown, some ten miles upstream from Boston. Others traveled down the shore and settled at spots that became known as Dorchester, Weymouth, and Hingham. Some joined Samuel Maverick on Noddles Island, and yet others stayed in Charlestown. Tellingly, all the early settlers chose islands, peninsulas, or at the very least the seashore as their homes: the islands offered relative safety from whatever dangers lurked on the mainland, and, more important, all these places allowed ready access to the only reliable means of travel, that by water.

The early English colonists in New England had a very ambivalent relationship with the coastal regions. On the one hand, the coastal waters were insecure, ever-shifting spaces where storms raged and shoals, sand spits, and submerged rocks threatened just below the waves. A number of early geographical names bear witness to the dread the region's topography inspired in the European newcomers: islands ominously called Great Misery or Norman's Woe, a shoal by the name of Devil's Back. Other places, such as Thacher Island, Avery's Ledge, or Collamore's Ledge, were all named for people who were shipwrecked on them, making them eternal reminders of the grim fate of unfortunate mariners. Early records of the colonies abound with tales of shipwrecks and boating disasters: John Winthrop relates in his journal how in December 1630 one "Richard Garrett, a shoemaker of Boston," and five others, including his daughter, "went towards Plymouth in a shallop, against the advice of his friends"; they proceeded down the South Shore and were almost at Plymouth when they got caught in a northwest gale that blew them out on the open water. They were able to land on Cape Cod and were assisted by local Natives, but it took days before a rescue party from Plymouth reached them and brought them to safety; by that time, four of them, including Garrett himself, had died of exposure.[10]

On the other hand, despite being inherently unsafe for the small vessels of the early settlers, the coastal waters offered the only way of communication between the colonies themselves and with the mother country. Pushing inland did not seem a particularly inviting prospect in comparison. Although several European newcomers commented on the open, almost parklike character of the landscape they first encountered in New England, this permeable environment was already changing by the time Boston was founded. The Natives, devastated by European diseases, no longer burned and cleared the forests, which were reverting to an impenetrable tangle of trees and underbrush.[11] Devoid of roads and once again overgrown with woods in which wild animals threatened, the mainland did not present a viable way of moving about during the first years of the colony. Water, on the other hand, provided for a comparatively easy way of moving people and goods from one place to another. Besides, compared to the many and largely unknown hazards of the coast, places like Boston Harbor soon became familiar and manageable: the number of dangers posed by navigating Boston Harbor was finite, and they could be learned about and monitored. By 1633 the harbor even had a pilot, John Gallop, who would come aboard incoming ships to guide them safely to the town's shores.[12] Between the threatening woods of the mainland and the shifting bars and violent storms of the coastal waters, Boston Harbor seemed a safe and sheltered space close to home.

The early settlements of the Boston area were quite literally built on water. Not only did most of the shoreline merge seamlessly with mudflats and salt marshes, but all of the towns had water access; some of them had more waterfront than uplands. Water washed the shores of Hingham and Charlestown, it encircled Samuel Maverick's small palisaded settlement on Noddles Island, and it seeped through the brown-green plains of mudflats stretching for miles from the shore of Dorchester into the bay at low tide. Boston was no different: the Shawmut peninsula, about two miles long and barely one mile across, was enclosed by water from three sides. In the north and west, it was separated from the mainland by the estuary of the Charles River. In the south, a deep cove ran between Boston and the peninsula of Dorchester Neck. In the east, the newly founded town faced the deep water of the harbor. Only in the southeast was the peninsula connected to the mainland by a narrow spit of land called the Neck. The Neck was low and marshy and would regularly flood during storms, making the Shawmut, with Boston on it, effectively an island. Even a large part of what passed as firm ground of the peninsula was marshy, and on the mainland beyond the Neck, this waterlogged environment continued in the swampy courses of Muddy River and Stony Brook. Thus, the early town of Boston stood almost completely surrounded by water.

Its location at the water's edge was one of the reasons the site was chosen for settlement in the first place. Boston's near-absolute separation from the mainland was seen as an asset by the contemporaries. William Wood, the author of a pamphlet designed to attract settlers to the colony, remarked in 1634 on the early Bostonians' fortunate situation, as "they are not troubled with three great annoyances, of Woolves, Rattle-snakes, and Musketoes."[13] Given their town's position, early Bostonians developed what historians Petra van Dam and John T. Cumbler have called an "amphibious culture": the harbor and its tributaries were for them a source of goods as well as an avenue to move them.[14] Fishing and fowling were an early means of sustenance for the colonists, who profited from the productive littoral ecosystem: John Winthrop recorded a "great store of eels and lobsters in the bay" in the summer of 1632, adding that "two or three boys have brought in a bushel of great eels at a time, and sixty great lobsters."[15] Besides keeping the annoyance of rattlesnakes at bay, waterways also acted as roads: boats and pinnaces of all kinds were a common possession among the settlers, who would travel by water from one town to another around the harbor and sometimes venture outside the harbor to visit fellow colonies along the shore. Water connected rather than separated: accessibility in the early days of Massachusetts meant having access to a waterway. As early as 1632, ferryboats plied busily between Boston, Charlestown, Winnisimmet, and Noddles Island; in November 1633, a certain Richard Browne was allowed by the Court of Assistants, the colony's governing body, "to keepe a ferry over Charles Ryver against his house," connecting Boston to Cambridge.[16] The way across the water was sometimes not only the most readily accessible way but also the only way: as Governor Winthrop noted in his journal, the winter of 1630 was so harsh that the Mystic River had frozen over, and people from Charlestown were unable to come to Boston for sermon.[17] This applied as much to moving around Boston and the other towns of Boston Harbor as to traveling between the English colonies in New England. Dependent on waterways, the early colonies all sat at the water's edge.

John Josselyn, a young Englishman who visited the New World in 1638, remarked on how empty the mainland seemed to him: "The Countrey all along as I sailed, being no other than a meer Wilderness, here and there by the Sea-side a few scattered plantations, with a few houses."[18]

The mainland was, despite Josselyn's impressions, of course anything but empty. By the time the Puritans started arriving in what they called New England, Native peoples had lived in the region for some ten thousand years. Big-game hunters at first, they kept adapting to their changing environment as the climate became warmer and the sea level rose in the aftermath of the last ice age.[19] In the centuries before the Europeans' arrival, they adopted agriculture, in addition to their established hunting-and-gathering practices, and led a combination of a sedentary and migratory existence designed to use the land's resources in an optimal way. Estimates suggest that approximately sixty to seventy thousand Natives lived in southern New England villages at the beginning of the seventeenth century. It was the contact with the Europeans that spelled doom for the Native culture: lacking immunity to European pathogens, an estimated 75 to 90 percent of the New England Indians fell victim to introduced diseases. Entire villages were wiped out, the Indians' economic and social life was disrupted, and large swaths of the land they had formerly lived on were abandoned.[20] Despite this catastrophe, Native Americans still peopled the land in which the colonists settled. In the first years after their arrival, contacts between the European newcomers and the Natives were frequent and most of them friendly, if cautious; the Native Americans' knowledge of the land and its resources made it possible for the Europeans to survive the first winters in the region's unfamiliarly harsh climate. The two groups had adapted to each other, established a complex network of relationships, and for a short time formed what historian James Drake has called their "own unique society."[21]

The connection the English colonists sought, however, was primarily to their fellow immigrants: not only were they on the same "civilizing" mission, attempting to turn the new territory into a version of England, only better and more virtuous, but some of them even knew each other from the streets of their hometowns back in Mother England.[22] In early colonial America, only the waterways provided this connection. Roads were virtually nonexistent, and travel over land was an arduous and insecure undertaking at best and a life-threatening one at worst. Even in the late seventeenth century, maps of New England and Virginia still showed a string of settlements along the Atlantic Coast, while leaving the interior largely empty. The English, Dutch, and French colonies along the Northwest Atlantic coast, rather than being defined by their common location on the North American landmass, were joined together by their shared access to the North Atlantic. They were what scholars of the Pacific insular culture have called a "sea of islands": places connected rather than divided by water. The waters of the North Atlantic functioned not as a barrier but rather as an "ocean highway."[23] More than settlements on the edge of a continent, the first European colonies in New England resembled an archipelago, a cluster of small islands tied together by the string of water.

Early modern European culture held some deeply ambiguous views about islands. On the one hand, they filled the role assigned to them by traditional European imagination: they were the seats of gods and lairs of monsters. On his voyage to New England in 1638, John Josselyn casually mentioned in his travel diary that one June day, in exceptionally foggy weather, their ship passed "an inchanted Island . . . but [we] could see nothing by reason of the mist: towards Sunset, when we were past the Island, it cleared up." Josselyn, who met the natural wonders of the Atlantic and of the New World with a bulwark of his day's prejudice, made it sound equally plausible that an "Island of Ice" they encountered a while later had "two or three Foxes, or Devils, skipping upon it."[24] Islands, not yet subjected to the sobering effects of triangulation and scientific study, were home to strange peoples and creatures known only from fables. They were universes contained in themselves, exceptional, mysterious, isolated. On the other hand, European islands had for centuries formed part of mainland cultures. Small islands around the massive one of Great Britain had been homes to fishing and farming communities. Islands of the Mediterranean had been stepping-stones for the expansion of European empires; those of the eastern Atlantic were subject to multinational colonization. Rather than leading a sleepy existence at the edge of the known world, Mediterranean islands often found themselves in the center of activity. Large landmasses assumed their position as central geographical categories only in the late sixteenth century.[25] In the early modern era, islands, far from disconnected, were tightly woven into the economic and political networks of the time, separated from the mainland neither in space nor in use.

"All the Ilelands within the Lymitts of this pattent"

In the amphibious environment of Boston Harbor, the Boston Harbor Islands were a natural part of the town of Boston from the start. The earliest colonists, like David Thompson or Samuel Maverick, had chosen them as sites for their homes because of the geographical advantages they offered. When the Puritans arrived in Boston Harbor, they no sooner started building their towns than they brought the islands under their jurisdictions and started using the resources they offered: in the early town, those were mainly timber and pasturage.

The Puritans had a utilitarian attitude toward New England's coastal islands since they first glimpsed their contours on the horizon. Francis Higginson, who came to New England aboard the *Talbot* in 1629, saw in the islands' "gay woods and high trees" signals of fertility of the new land, and as the ship passed down along the coast on its way to Salem, he thought it "wonderful to behold so many islands, replenished with thick wood and high trees, and many fair, green pastures."[26] In spite of seemingly sounding a note of pastoral romanticism, his remarks were in fact an accurate description of resources crucial for the success of the colony. Beside the colonists themselves, the Puritan ships also transported hundreds of sheep, goats, pigs, and cattle into the New World, and good pasturage was essential for the animals'—and their masters'—survival. High trees, on the other hand, were not merely majestic to look at but above all useful as sources of timber and ships' masts. In other words, the newcomers looked at the wooded islands of the Atlantic Coast and saw a land capable of sustaining them and their livestock.

John Winthrop and his companions moved from Charlestown to Boston mainly because of the latter's freshwater springs; however, with the good water source and the lack of wolves and "musketoes," the list of the site's immediate advantages was more or less exhausted. The Shawmut peninsula was rather small, surrounded by marshes, and the already limited amount of available land was further reduced by a number of steep hills that, though useful as lookout points, were less than advantageous as pastures, fields, or building ground.[27] Moreover, the peninsula had no trees. Animal fodder and wood for burning and building had to be fetched from elsewhere. According to William Wood, the Bostonians' "greatest wants be Wood, and Medow-ground, which never were in that place."[28] Elsewhere meant the mainland, but in the first place the islands of the bay, which were the readiest sources of both.[29]

When the settlers arrived in 1630, most of the islands were wooded, overgrown with typical New England species such as chestnuts and oaks.[30] Some records suggest that this may have been a rather recent development: the planters of Plymouth visited the future Boston bay in September 1621 on their way to the Massachusett people and reported that "most of the islands have been inhabited, some being cleared from end to end."[31] In fact, archaeological evidence suggests that most of the harbor islands had been used by the Natives for millennia, especially after rising sea levels had given Boston Harbor its current shape around five thousand years ago. Too small to support year-round settlement, the islands were most likely used as seasonal camps, particularly in the fall. Local Native people came to the islands to profit from the rich coastal environment and harvested soft-shell clams, oysters, and cod, as well as large mammals such as deer. Around the year 1000, they probably also established small farming hamlets at some of the larger islands close to the shore. By the time John Winthrop and his party arrived in the harbor, however, the islands were likely abandoned after the devastating 1616 epidemic that had killed most of the local Indians. The islands fell into disuse and were being slowly covered again by shrubs and trees.[32]

When the colonists arrived, the islands' accessibility and richness of resources soon attracted their attention. Each family in the colony was initially allotted a small piece of land to build a house on, a field in the as yet uninhabited portion of the Shawmut, and granted usage of the common pasture in the western corner of the peninsula near the Neck, which in the centuries to come would evolve into Boston Common. This area, however, was not enough to support a population, and the townspeople soon set their sights on the harbor islands.[33]

Most of the islands were not settled or even privately owned by the colonists in the first years of the settlement. In July 1631, one year after the arrival of the Boston colonists, the colony's Court of Assistants ordered that

> all the Ilelands within the Lymitts of this pattent vz. Conants Ileland Noddles Ileland Thompson Ileland togeath' with all other Ilelands within the lymitts of our pattent shalbe appropriated to publique benefit & vses, & to remaine in the power of the Governor and Assistants (for the time being,) to be lett & disposed of by them to helpe towards publique charges, & that noe person w'soeu' shall make any vse or benefitt of any of the said Ilelands, by putting on Cattell, felling wood raiseing slate etc. without leaue from the Governor & Assistants for the time being.[34]

The islands had what everyone in the colony needed and had to be treated accordingly. The Puritans, some of whom carried the idea of commonly owned and used land from Old England to the New, viewed the islands of the harbor as resources that were best taken care of when owned communally and used for the common good.[35]

They appropriated the islands as commons and used some of them for the benefit of the whole town, while leasing out others to the townspeople. Those islands were divided into roughly equal portions, so that everyone would have access to both upland pasturage and marshes at the water's edge where salt hay—an excellent winter fodder for cattle, as the colonists soon discovered—could be harvested.[36] Only on rare occasions were individuals allowed to use their own "private" islands: in 1632 Governor Winthrop was granted a large island across the channel from Boston, henceforth known as the Governor's Garden, or simply Governors Island, on the rather curious condition—given the prevailing local climate—of planting a vineyard on the island. Similarly, Samuel Maverick, the Anglican settler of Noddles Island, was granted his fair isle with an ordinance stipulating that Boston and Charlestown could still fetch wood from the southern portion of it.[37]

The first islands to be appropriated by Bostonians were naturally the larger ones and those closest to the town. In 1635 the Court of Assistants granted four of the harbor islands to Boston to use "forever."[38] Long, Deer, and Hog Islands were among the largest in the harbor, while the fourth one, Spectacle, so named because of its shape that vaguely resembled a pair of eyeglasses, was a midsize island that lay just south of the main shipping channel and was well accessible for boats coming from Boston. Some forty-two acres large at high tide, the island consisted of two glacial-till drumlins, connected by a low and narrow sand spit, the nosepiece of the imaginary spectacles. The smaller southern hill sloped rather gently toward the shore, but the higher northern drumlin rose in a steep cliff from the harbor waters. While deeper water on the island's western shore offered access for boats, tidal flats on the eastern side provided a suitable environment for a salt marsh and clam beds. Like most of the other islands of the bay, Spectacle offered the resources early Bostonians sought on the islands. Its two wooded drumlins were a source of timber and fuel. The marsh on its eastern shore supplied the townspeople with salt hay, and once the high ground had been cleared of trees, it offered pasturage for Bostonians' livestock.

In early Boston, the question of where cattle, sheep, and swine were supposed to feed was a rather troublesome one. The uninhabited portion of the Shawmut peninsula, where most of the townspeople's cattle usually grazed, had definitive shortcomings as pasture. Cattle from Dorchester habitually crossed the Neck and, disregarding the violation of jurisdiction, grazed on Boston's common pasture; the vacant part of the peninsula also doubled as field and vegetable garden, planted with corn, cabbages, and squash. The interaction between cattle and crops was not always a happy one: the resulting nuisance was apparently so common that in May 1631, the government of the colony found it necessary to decree that "all swine that are found in any mans corne shalbe forfeict to the publique" and the corn's owner compensated for damages.[39] Grazing cattle on the harbor islands presented a ready solution to the problem: not only were the cattle safe from wolves on the islands, but they could not wander off, which in turn made the townspeople's crops safe from the cattle.

Like the other islands, Spectacle started out as public land used primarily for harvesting wood. John Winthrop noted in his journal that in January 1638, "about thirty persons of Boston going out on a fair day to Spectacle Island to cut wood (the town being in great want thereof)" got caught in a snowstorm, and some of them were swept as far as the outer islands.[40] By 1640, however, the island commons was permuted into individual allotments of a few acres each, which were further transformed into permanent land possessions by the end of the decade.[41] This change was repeated on the other islands of the bay and reflected the development of Boston in the first two decades of its existence: the community experienced rapid growth, and land around town was becoming scarce. The townspeople were forced to resort to the islands not only for the resources of wood, fodder, and stone and gravel as previously but also for agricultural land. In the late 1630s, the need for land for planting seems to have been greater than the area of land available, leading to a speedy parceling and allocation of the islands.[42]

In the early days of Boston, Spectacle and the other islands filled the role of a seamless extension of the town's space. They supplied what Boston, in its rather confined location, lacked: pasturage for animals, timber and stone as building material, and wood for burning. Some of the islands' names, such as Hog, Sheep, or Slate Islands, reflected these early uses. Later, as the town grew larger, they partly made up for the scarcity of agricultural land around Boston. Although from today's point of view the islands' uses seem strictly rural, in the seventeenth century their position was much more ambiguous. Their uses were nonurban, but so were some of the activities that went on in the very streets of the early Boston. Seventeenth-century Boston was more a village than a town, and food production, resource extraction, and an unorganized waste disposal coexisted with its standing as the seat of government and principal port of the colony. It was a place where domestic animals roamed the streets and where residents had to be reminded by the town's selectmen not to "throw forth or lay any intralls of beast or fowles, or garbidg, or carion, or dead Dogs or Catts, or any other dead beast or stinkeing thing, in any hie way, or dich, or Common, within this neck of land of Boston." It was a place where cows grazed next to the houses of their owners and where there was hardly a well-to-do resident who did not have a barn, a stable, and a garden attached to his dwelling.[43] The division between urban and rural was very much blurred in the early town, making the islands a typical rather than unique part of the town. They were essential for the early Boston, supplying it with building materials and providing space for food production. Accordingly, they were brought under Boston's jurisdiction much sooner than most places on the mainland. In the early days, they were also more easily accessible than almost any other part of the town's surroundings, and although approaching them was not without its perils, as the wood-cutting party sent to Spectacle in the winter of 1638 learned to their chagrin, it was still preferable to finding one's way between the thick woods and impenetrable marshes of the mainland. Rather than apart, the Boston Harbor Islands were during the first two decades of the town's existence very much in the center of its economic activity.

Boston was by no means the only colonial town on the Eastern Seaboard that made use of adjacent islands in this way. The townships around Narragansett Bay used the bay's many islands for much the same purposes as the people of Boston used theirs. In the late 1630s, the Narragansett Bay colony's minister Roger Williams purchased several smaller islands in the bay and named them, in an unfathomable associative chain, Prudence, Patience, Hope, and Hog. As only the last island's name suggested, it was his intention to use the islands as swine pastures, capitalizing on their quality as "natural corrals," providing safe pastures for livestock while at the same time preventing them from damaging crops without the necessity of fencing. The even smaller Goat Island was, accordingly to its name, used for similar purposes, albeit for a different animal. In the mid-eighteenth century, Prudence Island hosted several farmsteads that made it a rural counterpart to the largely commerce-dependent towns that lined the shores of the bay.[44]

New York also made use of the many islands that filled its harbor (not to mention being built on one): mid-eighteenth-century visitors mention hay and barley harvesting on Governors Island, some eight hundred yards off the southern tip of Manhattan. Salem, the first important port of New England, used Bakers Island at the entrance of its harbor for farming and cattle grazing and made use even of the tiny Eagle Island for salt-hay mowing. Along the coast of Maine, there are until this day no fewer than eleven Hog Islands, with Calf and Cow Islands coming in a close second.[45] In early colonial New England, island farms and pastures were very much the norm among the coastal towns. The colonists regarded nearby islands as natural extensions of their settlements' space and used them accordingly. When their greatest need was agricultural land, they tilled, planted, and grazed their cattle on the islands. As their economy gradually shifted its focus to commerce, so, too, changed the uses of their insular appendices.

The Blessing of the Bay

When the Puritans first arrived in New England, they intended to establish a godly community that would be based on their religious doctrine while also ensuring their economic well-being. While most of their English predecessors in New England were sailors and adventurers whose main occupation was exploiting the rich fishing grounds of the North Atlantic coast and trading with the Natives, the Puritans' intentions and social background were different. They were often artisans and small landowning farmers, mostly coming from the English countryside. Many of them were prosperous: John Winthrop himself was a well-to-do country squire back home. They came to New England to change its supposed wilderness into fields, gardens, and orchards, seeing themselves as "planters" and their settlement as "plantations."[46] However, for those who settled in Boston, farming and husbandry were never a viable option. The Shawmut peninsula was too small to support more than a few dozen people, and as the town grew the amount of available land had become all but insufficient. The site possessed one important advantage, though, that more than made up for its inadequacies as a farming colony: the access to a large and sheltered natural harbor. William Wood summed up the town's prospects in 1634 with rare foresight: "These that live here upon their Cattle, must be constrayned to take Farmes in the Countrey, or else they cannot subsist; the place being too small to containe many, and fittest for such as can Trade into England, for such commodities as the Countrey wants, being the chiefe place for shipping and Merchandize."[47]

Boston Harbor was without a doubt one of the best places for safe anchorage and for establishing a seaport in Massachusetts, if not the whole of New England. When a party from Plymouth sailed into the bay in 1621, they declared that "better harbours for shipping cannot be than here are" and came back to Plymouth with "a good report of the place, wishing we had been seated there."[48] Their complaint might have had more to do with the notoriously rocky soil of Plymouth that afforded its colonists only a meager sustenance, but if their chief interest had been trade, their preference for Boston Harbor would have still been more than understandable. In Boston Harbor, geological forces that molded the New England coast had crafted a place superbly suited for shipping. Thirteen thousand years ago, what is now Boston Harbor was a wooded landscape dominated by rolling hills, winding streams, and deep-cut valleys. This landscape was created by glaciers of the last ice age pushing material before them and then receding, leaving behind hills of glacial till, or debris, known in geology as drumlins. Around five thousand years ago, sea levels rose rapidly, forming the present-day Boston Harbor: river valleys turned into deepwater channels, lower hills became underwater mounds, and the higher elevations emerged as the Boston Harbor Islands.[49] With the arrival of the Europeans, this half-submerged landscape came to be praised for its qualities as an excellent nature-made port facility. Vessels coming in from the outer bay would enter through a relatively narrow passage between two headlands that screened out much of the high winds and rough seas of the open ocean. After passing through this entrance, a spacious bay opened in front of them: early authors declared that "there is roome for the Anchorage of 500. Ships," which, given that the total area of Boston Harbor is approximately forty-seven square miles, was still grossly understated.[50] About 75 percent of the harbor had a depth of ten feet or less at mean low water, and this relative shallowness made land making and wharf building comparably easy. The harbor's physical disadvantages—the ubiquitous sand spits, underwater rock outcrops, and submarine mounds that made navigation in the harbor dangerous for large vessels—had become a serious problem only centuries later when oceangoing ships reached gargantuan proportions. The smaller vessels of the age of sail found their way around such obstacles more easily. Several deepwater trenches, the former river valleys, ran through the harbor, providing natural ship channels. The chief of them was the King's Roads (opportunely renamed President Roads after the Revolutionary War) that led through the harbor's northern part directly to Boston's shores.

The Boston Harbor Islands played an important part in making the harbor a safe and inviting environment for shipping. Richard Mather's joy at seeing "a great many of islands on every side" on reaching Boston Harbor after the frightful storm of 1635 was well founded: apart from the two headlands, it was the bay's many islands that shielded its waters from the onslaught of waves and calmed the rough winds coming from the east.[51] The rocky Brewsters guarded the harbor entrance, and the almost two dozen islands landward of Deer Island and Nantasket divided the harbor into several calm basins. Rising above the water masses, the islands provided easy orientation in the harbor for ships sailing to Boston. While the first map of the town of Boston, drawn by John Bonner, was printed in 1722, by that time a number of charts of Boston Harbor already existed, the oldest one dating back to 1688. This map, while showing the shoreline only as a thin red line and disregarding all that lay beyond that line, represented in minute detail all the harbor islands, the course and depth of the channels running between them, and also all the potential hazards lurking below the water surface: the tidal flats surrounding the islands, the long sand spit trailing behind Great Brewster Island like an animal's tail, the treacherous rock outcrops of Great and Little Faun. This early chart was commissioned by the royal governor of New England, Sir Edmund Andros, and its layout clearly showed the perspective of the colonial administration: it was interested less in the town of Boston itself than in the ways to get to it. In other words, the town's importance derived from its connection to the other colonial towns and to England. The waters of Boston Harbor were a highway leading to Boston, and the harbor islands were signposts and windbreaks along that highway.

The sea was an avenue for the arrival not only of friends and trading partners but also of potential enemies. Accordingly, one of the first maritime uses of the Boston Harbor Islands was defense: in 1634 the town of Boston had a primitive mud fort built on an island situated at the entrance to the inner harbor. This position, as Edward Johnson, a contemporary author, put it, was "very advantageous to

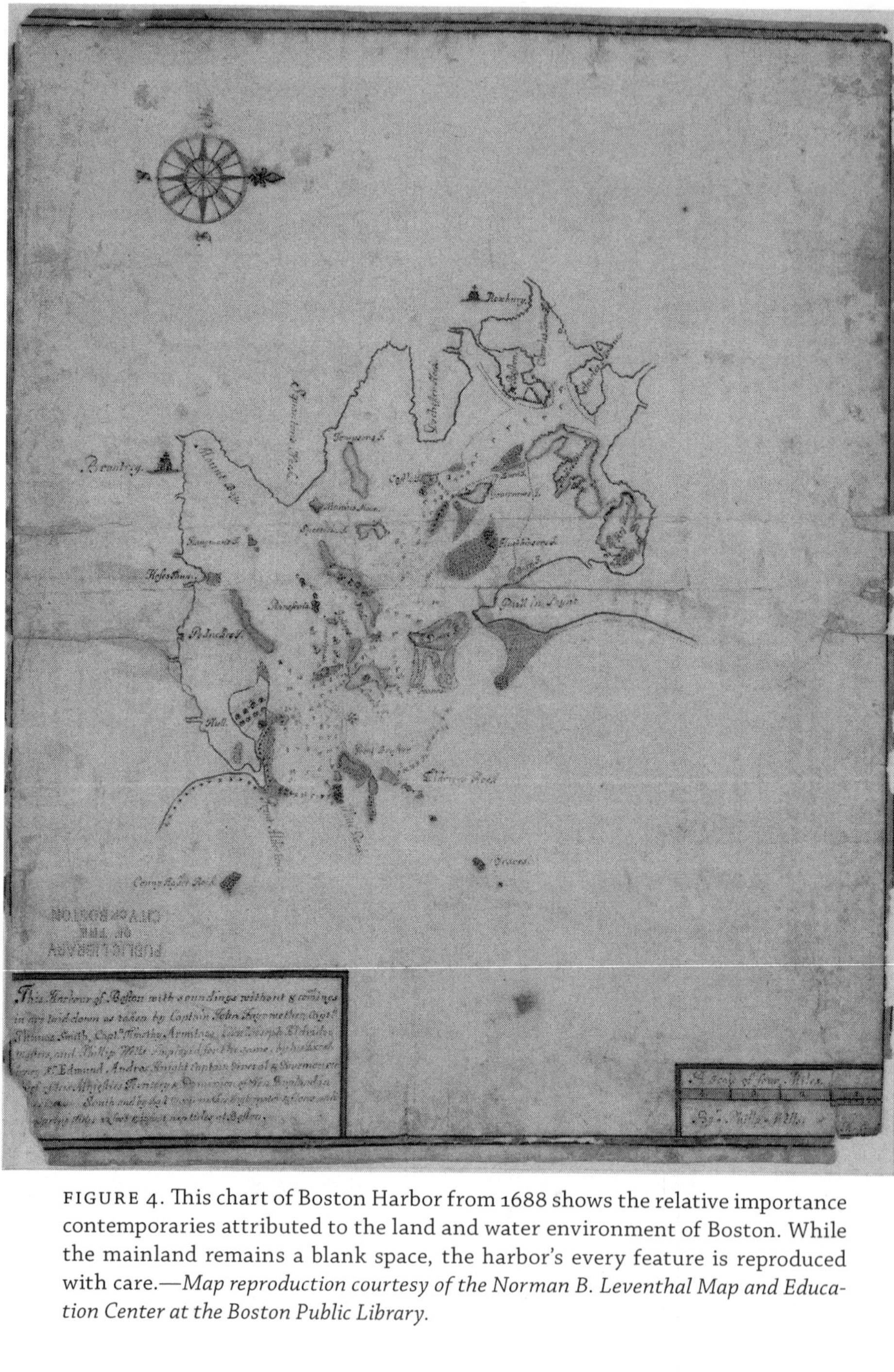

FIGURE 4. This chart of Boston Harbor from 1688 shows the relative importance contemporaries attributed to the land and water environment of Boston. While the mainland remains a blank space, the harbor's every feature is reproduced with care.—*Map reproduction courtesy of the Norman B. Leventhal Map and Education Center at the Boston Public Library.*

make many shot at such ships as shall offer to enter the Harbor without [Bostonians'] good leave and liking."[52] The original mud "castle" did not last long and was replaced first with a log building and later with a brick one.[53] It not only gave a name to its site, Castle Island, but was also the first in a long succession of military facilities on the Boston Harbor Islands.

Despite the Puritans' initial farming intentions, Boston was a seafaring town from the beginning. This development was partly fueled by sheer necessity: even when the islands and adjacent parts of the mainland were drawn into the process of food production, Boston's population, augmented by shiploads of new arrivals, still grew faster than the town's ability to feed itself.[54] Combined with its good harbor and the town's importance as seat of the colonial administration, it made Boston within a few years of its founding into the maritime hub of the Massachusetts colony. Soon, most of the goods sent to the colonies from England went through Boston.[55] English as well as Dutch ships that only a few years before would have landed in Salem or Plymouth now made their way to Boston, bringing new settlers, provisions, seeds of agricultural plants, and herds of cattle. Already in June 1634, John Winthrop remarked in his journal that "the last month arrived here fourteen great ships, and one at Salem." During its first years Boston was on the receiving end of the shipping traffic, accepting goods and people from the mother country and partly relying on these shipments for survival. However, only a year after the founding of the town, in June 1631, Boston witnessed the launching of a thirty-ton ship called the *Blessing of the Bay*, commissioned and owned by John Winthrop. The ship was built in Boston and, although decidedly unimpressive by even the standards of the day, it was Boston's first seagoing vessel and soon plied busily the coastal waters of the North Atlantic colonies, venturing as far south as Long Island and the then Dutch colony of New Amsterdam.[56] It did not remain Boston's only oceangoing ship for long: Matthew Cradock, a London Puritan merchant and shareholder in the Massachusetts Bay Company, established a shipbuilding business up the Mystic River in Medford that by 1633 was turning out ships twice as large as the *Blessing*.[57] The rather poor soil of eastern Massachusetts quickly persuaded Bostonians to focus their energies on farming a plain more promising than their rocky fields: that of the sea. The abundance of fish and other marine animals in the coastal waters of the seventeenth-century Northwest Atlantic was legendary: unlike their European counterparts, American fisheries had not been subject to centuries of intensive harvesting.[58] The coastal waters seemed to positively swarm with mackerel, herring, alewife, bass, and lobsters. Above all, they offered a plenitude of cod that, dressed, dried, and salted, became a staple of New England's fisheries and on whose scaly back the maritime economy of the Massachusetts colony rested. This was at least how it looked from Boston's statehouse, the seat of colonial government, where in the early eighteenth century a carved codfish has been installed as a constant reminder of the animal to which Boston owed its wealth.

The "sacred cod" was not the only reason Boston advanced to the principal port of the Northwest Atlantic seaboard. Neither was its fine harbor alone responsible for the town's ascent. Despite the frequent boasts of Boston's merchants and sea captains about having "one of the noblest harbors in the world," the qualities of Boston Harbor were in fact, although considerable, not entirely unique: the Canadian Maritimes were closer to Europe and had better access to the rich fishing grounds of the Grand Banks, Maine's harbors were equally good, and the harbors of Chesapeake Bay had the advantage of milder winters and ready access to agricultural produce. Furthermore, Boston Harbor lacked significant tributaries comparable with the St. Lawrence River, the Delaware, or the Hudson that would allow easy access to the continental interior. Despite these shortcomings, Boston, because of its position between the fishing of the North and agriculture of the South, developed into a hub of New England coastal shipping and later into a commercial center of trade with the Caribbean.[59]

It was the trade with the English holdings in the Caribbean that made Boston's early fortune. The West Indies plantation economy focused largely on sugar production and imported most of the needed foodstuffs and tools from elsewhere, including New England.[60] By the late 1640s, Boston tradesmen would ship timber and other wood products, dried codfish, meat, and livestock to the Caribbean islands and import sugar, molasses, and manufactured goods.[61] Boston's emerging merchant class also had stakes in the West Indies slave trade: many of them owned property in the Caribbean and traded and possessed African slaves. Boston's early settler Samuel Maverick was one of the first slave owners in New England: in 1638 his household reportedly included three enslaved Africans. The subsequent owners of Maverick's home at Noddles Island were involved in the slave trade well into the eighteenth century, making the island something of a local hub of the dubious business.[62]

The Caribbean was not the only source of the town's wealth: Bostonians would participate in the prosperous coastal trade that evolved between the colonies of New England and Virginia, their ships sailing up and down the coast loaded with tobacco, grains, and seal and beaver skins.[63] By the late 1640s, Boston had become the up-and-coming colonial center of shipbuilding and outfitting and evolved, over the short span of scarcely twenty years, from a community of would-be farmers into a bustling town of merchants, seamen, and artisans whose living came not from the land but from the sea.

Boston Looks Seaward

With the boom of their shipping trade, Bostonians turned their eyes to the sea both metaphorically and quite literally: it was not long before they set about shaping their harbor to enhance its qualities as a port. First, they started building wharves that would extend beyond the tidal flats surrounding the town, so that even larger vessels would be able to dock at them. Their efforts were supported by the colony's law: the colonial ordinance of 1641 stipulated that "the Proprietor of the land adjoyning shall have proprietie to the low water mark where the Sea doth not ebb above a hundred rods, and not more wheresoever it ebs farther."[64] With this law, Massachusetts, unlike other colonies (and indeed, unlike most present-day states), extended the property rights to coastal lands down to the low-water line, instead of the more commonly used high-water mark. The act was meant to encourage wharf building in the mostly shallow coastal waters of Massachusetts' harbors. Boston's merchants seized the opportunity: during the 1640s, more than twenty-five separate wharves were licensed to be built, and John Josselyn, when he visited Boston for a second time in 1663, observed that the town was "wharfed out with great industry and cost."[65] This sudden surge of activity changed the role the harbor space played for Boston. In the span of only a few decades, Boston Harbor added another important aspect to its function: the place that provided connection to the outer world and an avenue through which people and supplies arrived to Boston became also the primary site of the town's economic activity. It emerged as a place of labor, both extractive and productive, and began to adapt physically to its new role. Land making, wharf construction, and the building of docks were changing the contours of Boston's shoreline. The first places where this happened were the Town Cove, a deeply carved bay on the eastern shore of the Shawmut peninsula that directly faced the harbor, and the North End, a neighborhood situated at the Town Cove's northern shore, where most of those who made their living off the sea lived.[66] Soon, the original undulate shore of marshes and flats gave way to a cluster of wharves jutting into the deeper water of the cove. Changes occurred also in the parts not directly facing the harbor. Since these sites had no deep-water access, they were available for uses not associated with shipping. In the late 1640s, the deep cove between the North End and the then largely uninhabited part of the peninsula later known as the West End was dammed off, and several tide mills were erected on the resulting pond.[67] Seventeenth-century land making started what Boston's historian Walter Muir Whitehill has called "tinkering" with the town's shoreline. Although not comparable in scale with the wholesale remaking of the harbor's geography and ecology that occurred in the nineteenth and twentieth centuries, it was the beginning of a phenomenon that came to characterize much of Boston's historical interaction with its marine environment.[68]

Boston's new commercial identity manifested itself not only on the waterfront but also on the harbor islands. As agricultural land lost its initial importance for Bostonians' subsistence, the scramble for insular land allotments was over: most of the islands became private property of a few individuals and were mostly used for grazing cattle.[69] The islands' immediate environment, however, was Boston's increasingly busy harbor: their location in the beating heart of Boston's maritime commerce made it only a question of time before their resources were taken hold of and used for the purposes of Boston's prosperous shipping. First, the islands were still mostly wooded. The island trees, as the shipbuilders of Boston soon discovered, could be equally well processed into ships' planking as when they had been formerly used to support the roofs of Boston dwellings. In 1634 William Wood pointed out the islands' arboreal vegetation to seamen who, "having spent their old store of Wood and Water, may haue fresh supplies from the adjacent Ilands, with good timber to repaire their weather-beaten Ships: Here likewise may be had Masts or Yards, being store of such Trees as are usefull for the same purpose."[70] Similarly, when Bostonians built their first wharves, they reached out to this familiar resource. Seventeenth-century wharves consisted mostly of a wooden platform or crib that was sunk to the bottom and into which fill was then deposited. As Boston itself had no trees, the likeliest sources of timber were the mainland and the harbor islands.[71]

Once the trees were gone, the town turned to another resource the islands offered and the shipping traffic needed. The ships of the period would carry stones or gravel in their bellies as ballast; they would dispose of them or take them on according to current need at the various ports they visited. Some of the ships that came to Boston unburdened themselves of their ballast stones there: this practice accounts for European flints found in Boston's made land.[72] Other ships took on ballast while in Boston, and more often than not, they helped themselves to gravel and pebbles from the readiest source, the harbor islands. The islands' owners recognized a business opportunity when they saw one: many of them started actively selling their islands for ship ballast. In the early nineteenth century, at the port's zenith, they harvested as much as 150,000 tons of stone and gravel from the islands each year.[73]

This long-term, large-scale exploitation understandably left deep scars in the islands' shape and ecology. As the years went by, the more the port of Boston expanded, the more the harbor islands shrunk. Whereas renderings from the early eighteenth century, such as the 1708 edition of *The English Pilot*, depict the Boston Harbor Islands as largely wooded and record Slate Island or Nixes Mate as substantial islands of several acres, 1770s views present the observer with a picture of an insular landscape virtually denuded of trees and dominated by severely eroding cliffs. Decades of deforestation and gravel harvesting had taken their toll. They left many of the islands exposed to wave action and erosion, causing them to slide gradually into the harbor waters, and—in the case of Deer Island—right into the main shipping channel. The island, one of the harbor's principal headlands, had sustained a tremendous loss of an estimated sixty acres of upland over the course of the eighteenth century—and its losses were the rule rather than an exception. Nixes Mate, an island of twelve acres in the 1630s, had lost half its area by 1769 and by the end of the eighteenth century had been apparently reduced to little more than a shoal, visible only at low tide.[74]

Amid the ballast quarrying and mast cutting, some islands retained their older agricultural function. Some of the larger islands had been transformed into farmsteads: Joseph des Barres's 1777 surveys of Boston Harbor show many of the islands as pastoral landscapes, a combination of meadows, orchards, and stands of trees. Spectacle Island was one of the islands that accommodated both the old uses and the new. The island was, since 1684, the private property of Samuel Bill, a Boston resident whose father possessed one of the original fifteen grants on the island.

Bill Jr. bought the rights to the whole island from Josiah, son of Wampatuck, a Massachusett sachem, and established a farm on it: a 1729 deed in which Samuel Bill sold the island to his brother Richard listed a "dwelling house, barn, buildings, fences, and trees" on the northern drumlin.[75] Farming and pasturage use continued for the whole of the eighteenth century and beyond, yet apparently did not prevent the island from being plundered for ballast material at the same time. In 1822 Joshua Henshaw, the then proprietor of Spectacle, threatened prosecution in a series of newspaper notices to those who illicitly removed gravel and stone from the island and sold it as ship ballast.[76] Thus, the island, while functioning as part of Boston's rural hinterland, was at the same time scoured for resources useful to shipping and drawn into the workings of the town's maritime economy.

During the remainder of the seventeenth century and over the whole of the eighteenth, Boston's economy continued to grow. Boston merchants kept increasing their radius: what started as trading along the coast and shipping goods to and from the West Indies and England expanded during the mid-eighteenth century to include commercial contacts with the Mediterranean and the Baltic and toward the end of the century also the prestigious China trade. Boston, apart from selling the colony's products such as lumber and cod, served also as a trading hub for produce from other regions, making them its "alternative hinterlands." Thus, both fish from the Grand Banks and cotton from Alabama would move through Boston.[77] Boston merchants benefited from trade with the Baltic, buying Russian hemp, iron, and linen. They also made inroads in the Mediterranean, importing oil, fruit, and wine. Between 1714 and 1717, some 1,300 vessels left Boston carrying dried codfish, whale bone and oil, pickled mackerel, lumber, and also New England rum made from Caribbean molasses.[78] Boston was governed by men who made their fortunes in overseas trade: in 1687 five of the town's nine selectmen were sea captains, and more than a half of the town's wealth was controlled by those engaged in trading with England and the West Indies, building ships, and

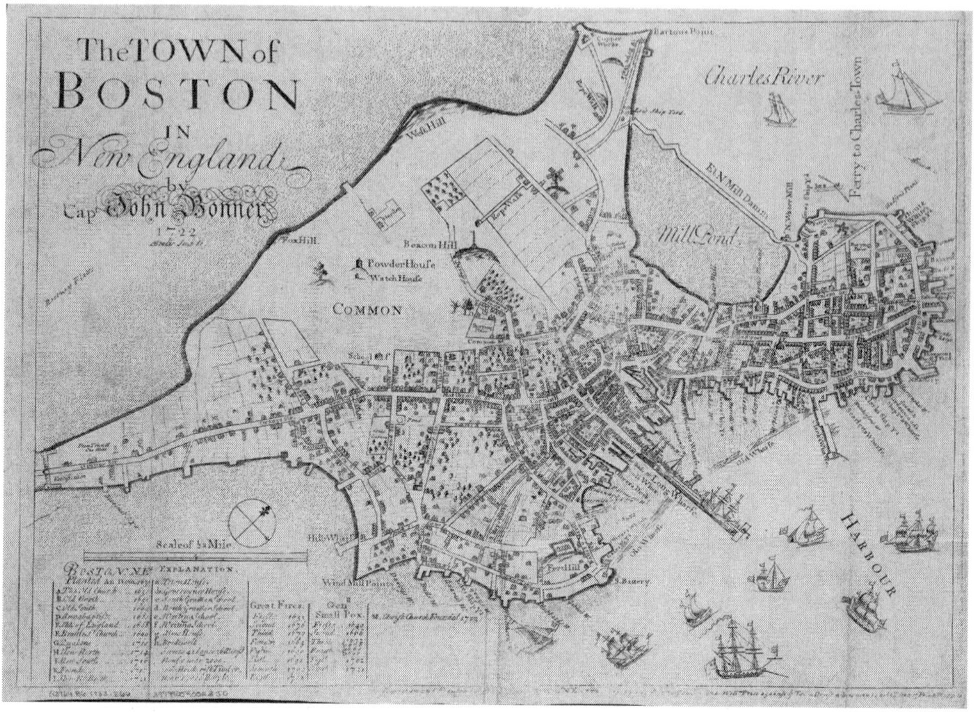

FIGURE 5. A 1722 map of Boston by John Bonner. The seafaring town of Boston was crowded on the harbor-facing side of the Shawmut peninsula. The town's most prestigious piece of waterfront development, Long Wharf, can be seen jutting out in the middle of the Town Cove. —*Map reproduction courtesy of the Norman B. Leventhal Map and Education Center at the Boston Public Library.*

distilling.[79] Boston was the most active colonial seaport not only in commerce but also in shipbuilding: between 1674 and 1714, 437 ships were built in Boston's shipyards, and products associated with shipbuilding, such as masts, boards, staves, sails, and ropes, made up a large part of the town's exports.[80]

The near-absolute dominance of seaborne commerce in Boston's economic life was distinctly reflected in the town's spatial organization. The famous map of Boston drawn by Captain John Bonner, first published in 1722, is perhaps the most powerful testimony to this maritime orientation. Early-eighteenth-century Boston is huddled on the shore of the Town Cove, facing the harbor. The busiest and most densely built-up neighborhoods on the map are those closest to the waterfront. Most of the streets lead down to the water's edge and do not even stop there: they continue into the harbor in the form of numerous wharves (more than forty can be counted on the map), blurring the division between land and sea. Even those areas without access to the waterfront are devoted to shipping and trade: more than a dozen shipyards and six ropewalks—low, elongated structures where ship ropes were made—are dispersed between the townhouses, in addition to workshops manufacturing sailcloth and distilleries processing Caribbean molasses, located conveniently close to wharves. The harbor waters are alive with carefully rendered ships, their precise detail in stark contrast with the rough outlines reserved for the part of the peninsula facing the mudflats of the Charles River. Anyone arriving aboard one of the sloops, brigs, and snows of the era would be likely to land at the proudest piece of Boston's eighteenth-century waterfront development: Long Wharf, built in 1710 by Captain Oliver Noyes and protruding one-third of a mile into the harbor.[81] The stately wharf more than anything else resembled a street casually extended into the water, complete with shops and warehouses on it.[82] Once the passengers disembarked, a majestic prospect opened itself in front of them: the wharf merged seamlessly into King Street, the town's central thoroughfare that led up to the statehouse, the seat of colonial government. This arrangement sent a clear message about how Boston understood itself and how it wished to be understood by visitors: as a seafaring town par excellence turned toward the outer world, a place with stronger ties to London thousands of miles away than to its neighbor Roxbury on the mainland. Boston Harbor and Long Wharf were from this perspective the only proper entrance to the town: the wharf was a symbol of the town's orientation toward its harbor and the network of trade connections that lay beyond and, at the same time, "the obvious avenue to Boston from the part of the world that really mattered."[83]

There was, naturally, also the part of the world that did not really matter. If space can be thought of in terms of the front and back of the human body, with the front being the more prominent and lively, and if such values can be detected in urban design, then colonial Boston definitely possessed a very distinct front and back.[84] Those approaching the town by sea were entering through its front gate. The waterfront was busy, even crowded, and all of colonial Boston's representative buildings were found there. The approach to Boston from the mainland over the Boston Neck felt, on the other hand, like slipping in through the back door. The Neck offered a "bleak, unpleasant road" that led the traveler past a few scattered houses, cows grazing on the marshy meadows, and the town gallows.[85] The spatial values favoring the harbor were apparent from the names bestowed on streets and topographical features: three streets in the shipping-oriented North End went by the names of Fore, Middle, and Back Streets, relative to their distance from the waterfront. A large, muddy cove on the western side of the peninsula, facing away from the harbor, was tellingly known as Back Bay.

Just as the wharves, the outstretched arms expanding Boston into the water, complicated the notion of a clear division between land and sea, so, too, did the harbor islands problematize the dividing line between Boston and its harbor. At first sight, they were clearly separated from the town by the intervening stretch of water. Yet in early Boston, water did not necessarily act as an interval between places but rather as a connecting medium. Especially in early-seventeenth-century Boston, the islands were not the remotest part of town but rather the most accessible. The islands were part of the littoral zone, a tremendously productive ecosystem that generously supplied the settlers with salt hay for their animals and with fowl, fish, and shellfish for their own consumption.[86] If anything, it was the mainland that seemed hostile, empty, and remote. Stemming at first from the limited amount of land and resources available at Boston's original site, and later from the demands of the town's maritime economy, the islands were indispensable for the functioning of the town: they were its fields, forests, meadows, and quarries. Their very presence made Boston Harbor what it was, an excellent natural shipping facility. Far from being separated or isolated, the islands of Boston Harbor were a natural extension of the town's space. The way their use, their ecology, and their very shape changed over time reflected Boston's development from a rural settlement into a town relying on shipping commerce for its prosperity. They were no different from Boston's other immediate surroundings: Bostonians fetched their wood and grazed their cattle both on their small peninsula and on Spectacle Island; they dug for gravel both on Nixes Mate and on the Trimountain. It was not until the mid-nineteenth century and Boston's emergence as a modern metropolis with sanitation problems to solve that the islands were cast in a new role: that of the city's dumping ground.

During its first century, Boston's spatial pattern remained much the same. At first sight, the 1722 Bonner map looks like an unfinished painting. The southeastern half of the Boston peninsula presents a precisely worked-out maze of streets, houses, plots behind the houses, and wharves in front of them; the western portion remains largely blank. Apart from lacking deepwater access, this was also the part of the peninsula dominated by several steep hills that presented a natural barrier to the town's expansion. Colonial Boston remained squeezed on a narrow strip of land, facing the harbor and leaning against the slopes of the Trimountain. Up to the late eighteenth century, the area of the West End, the Neck, and Boston Common was the undisputed backyard of Boston, a sparsely settled region that was home to the "common nuisances": institutions and businesses undesirable in the more fashionable or densely populated neighborhoods. The West End was the site of a few solitary ropewalks, cumbersome and highly flammable structures that were only grudgingly tolerated in the more populous districts. The vicinity of the common was seen as fit for a haphazard assembly of a powder house, a granary, and a burial ground. Because of their relative inaccessibility, hills were considered urban fringes and assigned uses undesirable elsewhere.[87] The Trimountain, a dominant topographical feature of early Boston, housed the unappealing ensemble of the town's jail, almshouse, and workhouse.

As the town grew, however, its fringes moved, too. William Price's 1769 update of the Bonner map shows the common increasingly encircled by newly laid-out streets. Toward the end of the eighteenth century, the changes accelerated. The 1795 construction of the new statehouse on Beacon Hill, designed by Boston's star architect Charles Bulfinch, made the elevation a desirable and genteel part of town and helped transform the common from a place where the townspeople were just as likely to take a pleasure stroll as to beat their carpets or graze their cows to a prime urban site of leisure.[88] The hitherto unobjectionable location of the workhouse and almshouse on the northeast side of the common became a problem due to their sudden close proximity to the newly built elite dwellings on Beacon Hill. In 1795 both institutions had to move to an even more remote part of town: Barton's Point, the westernmost tip of the still largely unsettled West End.[89] Boston's spatial organization was changing rapidly, as the growing town saw its previously "unsavory" fringes transformed into neighborhoods of its merchant elite. The "nuisances" had to make way; soon, there was hardly a place for them on the confined Boston peninsula. However, just like its commercial facilities, Boston's fringe did not stop at the water's edge. The next destination of the undesirable institutions and businesses was a region that gradually came to be regarded as the town's ultimate fringe: the harbor islands.

2

THE ULTIMATE SINK

The Harbor and the Islands as Boston's Dumping Ground

The summer of 1873 was not a good time to be in Boston. To be sure, the sun glittered on the golden dome of the statehouse, the hallowed halls of Harvard across the river emanated an atmosphere of accumulated wisdom, and the Great Elm on the Common stood just as tall as in any other year. By night, however, the Athens of America was enveloped in an atmosphere that had little to do with dignity or learning. The residents of many of the city's neighborhoods were repeatedly roused from their sleep by a broad variety of odors, coming from different sources and in different degrees of intensity, but all of them revolting. Some of them were the rotten-egg smell of decomposing sewage, others the sickeningly sweet odor of burning animal flesh. They permeated dwellings and clothes and filled the lungs. They made the afflicted population despair—and petition by the hundreds to the city's recently established Board of Health. The new public health agency was eager to demonstrate its usefulness and lost no time in responding to the complaints. After registering the Stony Brook sewer and the exposed mud of Back Bay as sources of only a minor portion of the foul smells, the board members, accompanied by the city physician, decided to investigate in the field. On the night of June 18, they set out in the opposite direction of the wind and soon arrived at one of the basins of Millers River between Cambridge and Charlestown. The banks of Millers River, a small tributary of the estuarine Charles River, were by the 1870s home to several meatpacking companies and fat-rendering establishments that dumped their waste into the slow-moving stream, basically turning it into an open sewer combined with a trash heap.[1] Sure enough, once the Board of Health members reached Millers River on their nocturnal foray, one of them distinctly recognized the sickening odor of cooking putrid meat that had kept him up half the previous night. Referring the matter to the board's state-level counterpart, the chairman of the city board concluded mournfully: "Many people sought better air on sea or shore, or among the hills; but there were many of us left who will long remember the summer of 1873 for its success in the dissemination of the most dreadful odors that have ever afflicted us."[2]

Memorable as the events of 1873 were, they were far from a singular occurrence. Only five years later, the summer of 1878 again aspired to the title of Boston's worst-ever smell season. The sewer outlets along the waterfront, as well as the stagnant, putrid waters of Stony Brook, Millers River, and South Bay, and several bone-boiling and grease- and tallow-rendering establishments in Boston and Cambridge collectively gave off a stench strong enough to "arouse the sleeping, terrify the weak, and nauseate and exasperate nearly everybody."[3] In an inevitably inadequate attempt to translate the olfactory into the visual, the Board of Health produced a one-of-a-kind map showing the sources of offensive odors around the city. Marked in accusing red, major offenders had been pinned down all along the city's waterfront and beyond: a guilty string of red dots lined the coastline, signifying sewer outlets; these were interspersed with an occasional crimson patch of sewage-infused mudflats. Taken at face value, the map represented an effort of municipal authorities to identify, explain, and ultimately contain the worst sources of stench within the city. At the same time, it virtually breathes the indignation of both the citizen petitioners and the board members about the nuisances depicted. Their outrage was, however, only partly directed at the simple fact that this should be happening at all; it was in no small part owing to the fact that this should be happening *in* Boston.

Smells are difficult to grasp. They are elusive and transient, and we lack a proper vocabulary to capture their quality.[4] Even

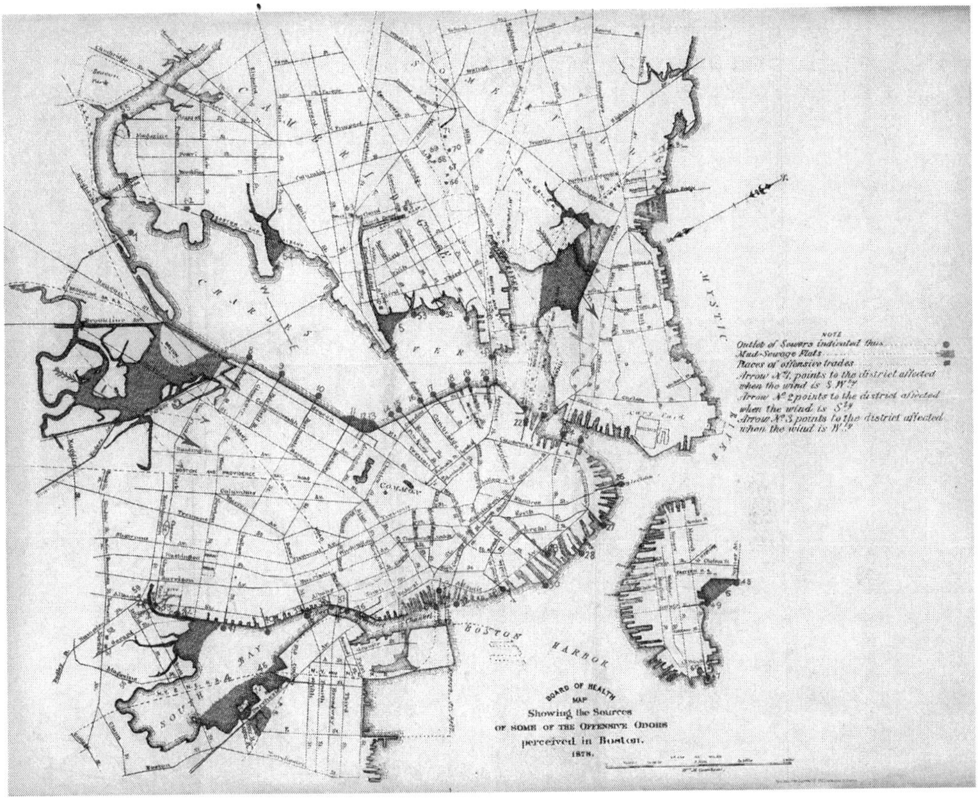

FIGURE 6. This 1878 map produced by the Boston Board of Health is a rare attempt to visualize the sources of "offensive odors" that plagued Boston at the time. —*Courtesy of Boston City Archives.*

nineteenth-century Bostonians, for whom odors played a much more important role than for us, used a limited set of adjectives to describe them. Odors that plagued the city were "vile, offensive, intolerable, alarming, dreadful, noxious, foul, objectionable, disagreeable, unwholesome, noisome and injurious," and simply "bad."[5] These words all related not to the quality of the smells themselves but to the effect they had on the people who perceived them. This was their most important feature: they were a nuisance that the urban population reacted against.

Bad odors in a nineteenth-century city were no mere annoyance: they had a far more ominous meaning. Miasmatic theory, the leading medical theory of the day, postulated that spread of diseases was caused by miasmas, poisonous vapors hovering in places with foul air and inadequate ventilation. The theory gave rise to the idea of the sanitary city in the 1840s and made places like Millers River seem like hotbeds of disease.[6] Such sites needed to be eliminated or at least removed as far from human habitation as possible. After 1850, for the first time in its history, Boston had both the will and the means to do so.

A Growing Boston

The second half of the nineteenth century was a time of sweeping changes for Boston. Between 1800 and 1900, Boston's population doubled every twenty years, transforming the relatively close-knit community of thirty-four thousand in 1810 into an urban behemoth of nearly a half million in 1890. At the end of the century, Boston found itself with a population more than twenty times larger than at its beginning and was bursting at the seams. The original peninsula, while able to house a population of several thousand, proved too small when this figure soared into six-digit numbers. To be sure, Boston continued its centuries-old practice of enlarging its territorial footprint by filling tidal flats and "wharfing out" into the harbor. In the course of the nineteenth century, the vast expanse of Dorchester flats was largely filled, creating the new neighborhood of South Boston.[7] These coastal developments, however, were slowed down in midcentury by the maritime community's concerns about the "health" of the harbor, meaning its continued suitability for shipping. Since at least the 1830s, the harbor appeared to be shoaling in many places, and a succession of harbor commissions was established to deal with the perceived threat to the city's maritime commerce. Most prominently, the United States Commissioners on Boston Harbor studied the problem in the late 1850s and, in a series of reports published between 1860 and 1866, came to blame the extensive land making for the harbor's apparent deterioration. Land making, so the theory went, disrupted the flow of water in the harbor and the workings of tidal scour that swept the channels free of sediment. Anxious to keep the harbor in working order for the port of Boston to retain its position as the nation's leading maritime gateway second only to New York, the harbor commissioners proposed a series of remedies. The most prominent of them was drawing a line along Boston's waterfront, beyond which no land making would be allowed. Although the line proved quite movable in the coming years and a significant amount of land making was tolerated, especially if compensated for by dredging elsewhere, still the commissioners' line imposed substantial restrictions on the city's expansion into the harbor.[8]

Boston's hunger for space thus turned in the opposite direction. Since the introduction of the horse-drawn streetcar in the 1850s, Boston spilled over rapidly onto the mainland, attaching the formerly independent communities of Dorchester, Roxbury, or West Roxbury to its ever-growing body and making them into "streetcar suburbs." Whereas in 1850 settlement mostly concentrated within a two-mile radius from Boston's city hall in the center of the downtown district, by the 1890s this distance increased to at least six miles.[9] Annexation of neighboring communities and modern means of transportation, mainly the streetcar—horse-drawn at first and later electrified—transformed the relatively compact commercial city of the first half of the century into a sprawling metropolis within only fifty years.

Territorial growth and comparatively easy commutes brought about the advent of a "city of work" separated spatially from the "city of homes."[10] Replacing the multifunctional pedestrian town of the past, new territorial units developed within the city, often devoted to a single purpose: commerce, industry, or residence. One of the side effects of this diversification was the creation of a hierarchy of places more pronounced than ever before. The newly constructed residential district of Back Bay with its handsome three-story brownstones and tree-lined parkways catered to the needs and tastes of the city's merchant and industrialist elite, while the suddenly accessible suburbs turned into relatively homogenous communities of the middle class. Meanwhile, the old waterfront district became home to thousands of people who came to Boston in subsequent waves of immigration from Ireland, Italy, and Russia.[11] Tenement houses went up in the "crooked and narrow streets" of the former heart of the town of Boston, to use a phrase famously coined by historian Annie Haven Thwing. The city's social stratification was increasingly being expressed in spatial terms, creating a system of clearly defined neighborhoods inhabited by people of similar social standing and ethnic background. In a much-quoted passage from John P. Marquand's novel *The Late George Apley*, a satire of Boston's upper class, the hero's father, who has just purchased a house in the supposedly "respectable" neighborhood of South End, observes one day a neighbor in his shirtsleeves on the steps of a house across the street. Alarmed by this display of wanting manners and correctly deducing the imminent social decline of the neighborhood, he sells the house the next day and moves the family to the more securely dignified Back Bay.[12]

In a world that strived for a clean separation of social classes and land uses within the city, and at the same time had both the space and the transportation means at its disposal to make this separation reality, close proximity of "high" and "low" uses was less and less tolerable. Since the beginning of the nineteenth century, the so-called nuisances and offensive trades were being pushed first to the fringes of the Boston peninsula and then altogether out of town. Proprietors of tallow works, slaughterhouses, and fat-rendering plants found in the 1870s that their businesses were no longer welcome within the city limits. In 1882 the firm C. F. Bowker & Co. petitioned the Boston Board of Health for permission to open a rendering establishment on Kemble Street near South Bay. South Bay, a remnant of the once-large marshy cove south of Boston Neck, was by the 1880s reduced, similarly to Millers River, to a sliver of stagnant, notoriously polluted water, its shores lined by run-down warehouses and railroad yards. It was a location for a rendering plant that only a few decades before would have raised few eyebrows. In the early 1880s, however, the proposal drew a large crowd of opponents of the scheme to the board's public hearing. The permission was eventually rejected, while the board commented: "It is hoped that the time is not far distant when all business of this nature will be conducted outside the limits of the city."[13]

It was not just the tallow works, however: public institutions housing the socially outcast—poorhouses, workhouses, and prisons—were, too, all assigned a vague yet at the same time perfectly clear place: as far from the upper-class neighborhoods as possible. The city's poorhouse, after having been pushed around the original peninsula in the 1790s to give way to new residential development, was in 1822 moved to South Boston, a new neighborhood across South Bay from Boston proper, so as not to take up valuable land and disturb the genteel neighborhoods of the city. It was soon joined there by a prison for petty criminals and a reformatory for juvenile offenders.[14] The ensemble of public institutions that received Boston's poor, criminal, and mentally ill in South Boston grew, prompting its residents to complain in 1847 that Boston was turning their peninsula into a dumping ground.[15] By the 1870s, even South Boston's population soared, and its complaints were becoming louder. South Boston residents started to organize into citizen associations that would promote and protect their interests. One of the most vocal of these organizations was the City Point Mutual Protection Association, representing the residents of an increasingly popular neighborhood at South Boston's southernmost tip that prided itself in its commanding views of the harbor and pleasant sea breezes. In 1873 the association issued a statement in which it demanded of the city government "the removal of public institutions from South Boston to a place of greater convenience and where their existence will not depreciate private property."[16] Just as Boston's city limits were on the move, so was its fringe.

Harboring Disease

In the latter part of the nineteenth century, Boston was quickly running out of places to put its nuisances, its offensive trades, and its undesirable public institutions. As the settlement grew denser and the population more outspoken, the siting of unwelcome businesses and institutions was becoming increasingly difficult. There was, however, still one wedge of the city's space that seemed underutilized and perfectly suited for such a purpose: the harbor islands.

The Boston Harbor Islands' position vis-à-vis Boston had changed significantly since colonial times. In the amphibious town of the seventeenth century, the islands formed stops along the "ocean highway" and were among the most visible and accessible parts of town. In the nineteenth-century city that sought to solve its inherent lack of space by expanding landward and that relied increasingly on land instead of waterborne transportation, the islands were gradually becoming the urban fringe.[17] Also, the foundation on which Boston's wealth rested had shifted significantly. The town's merchant class, whose business was disrupted first by the siege during the Revolutionary War and even more profoundly by the War of 1812, reoriented its focus from the shipping trade toward manufacturing.[18] In the 1820s, Boston's economic elite no longer looked eastward toward the sea but rather in the interior toward the manufacturing towns of Lowell, Waltham, or Worcester, where textile and leather factories sprang up. Combined with the residential boom of the suburbs, this development meant that in the second half of the nineteenth century, the city's waterfront and its harbor islands were being increasingly regarded as places most fit to receive trades and institutions that were unwelcome elsewhere.

The idea of the harbor islands as places apart was not entirely new. Since Boston's founding, the town had used not only the islands' proximity to the mainland but also their physical separation from the settlement on the shore. In the late seventeenth century, this ambiguous view of the harbor islands came to be articulated in sinister terms. In the fall of 1675, during the complex and violent conflict known as King Philip's War, Massachusetts Bay colony officials rounded up hundreds of Christian Natives from eastern Massachusetts's "praying towns" and forcefully removed them to Deer Island. Deer Island, although attached to the mainland today, was then still an actual island, and its position at the entrance of the harbor exposed it to the violence of harsh seas and winter storms. Lacking adequate shelter or food sources, the island soon became a dreaded prison site, a bleak exile that isolated the captives from both the English and their own communities. Before they were released again in the spring of 1676, an unknown number of the Native prisoners had died from exposure and starvation.[19] The choice of the site as in fact a prison camp was telling: Deer Island was apparently seen by the English colonists as isolated yet at the same time close enough for them to exercise control.

Since the 1640s, the islands, in their role as the town's outposts in the harbor, also served as ad hoc quarantines for incoming ships. Throughout the seventeenth century, ships suspected of carrying smallpox and other infectious diseases were detained on various islands along the main ship channel. In the early eighteenth century, the practice of isolating the sick on the islands of the harbor started to include not only ship crews and passengers but also Boston's own residents. So, in 1702, a widow Page and her family came down with smallpox and were removed from Boston to Apple Island. At first, several mainland sites were also used for isolating contagious patients. One such place was Fort Hill, an elevation in the southern portion of the Boston peninsula, somewhat removed from the main settlement. However, by the 1710s even this site was considered too central by Boston's selectmen, and a new, more remote, one was sought.[20] In this case, the harbor islands' sheer physical distance from what was by then considered "Boston proper" and the boundaries of which were more or less defined by the coastline of the Shawmut peninsula was equally important as their connection to the town. As would so often be the case in Boston's history, the islands' main asset was their ambiguous position in relation to the town: they were remote enough to offer a sense of separation from unpopular uses, yet at the same time close enough to allow easy access and enable the townspeople to control the goings-on on the islands. In 1717 Boston's selectmen established the town's official quarantine on Spectacle Island and directed all ships that entered the harbor with a contagious disease aboard to anchor there. In 1721 the ship *Seahorse*, coming from the West Indies and carrying, besides molasses, also the smallpox virus, was ordered to the island quarantine.[21] Unfortunately, this was not before several of the ship's sailors went ashore in Boston and spread the disease in the town. The *Seahorse*'s crew was soon joined on Spectacle by smallpox victims from Boston, making the island the town's official sick bay. In 1737 the Massachusetts General Court and Boston's selectmen moved the quarantine hospital to another, even more remote, insular site at Rainsford Island. From there it wandered in the 1860s to Gallops Island, closer to the ship channel and farther from Boston, cementing the understanding of the harbor islands as isolated and bleak and of the harbor waters as an interval between places rather than a connecting medium.

The use of islands as nature-made quarantines based on the simple principle of physical separation was nothing unusual in eighteenth- and nineteenth-century America. Fear of smallpox brought by ships led New York to establish a quarantine station in 1738 on Bedloe's Island, the future site of the Statue of Liberty. Unlike in the latter case, it was chosen not because of its central position on the sea approach to New York but rather for its convenient distance from New York's shore. Two decades earlier, a "pest house" was built on Coasters Harbor Island in Narragansett Bay.[22] A 1782 map of the harbor of Portsmouth, New Hampshire, shows Shapleigh Island in the harbor being used for the same purpose. Siting of medical facilities that were perceived as hazardous away from population centers on islands had become quite common in the eighteenth century. Sometimes, local residents did not consider even the supposedly remote islands to be remote enough. In 1771 when experiments with inoculation as a method of preventing smallpox were in full swing, Samuel Gelston, a noted practitioner of the new method, set up an inoculation hospital at Gravelly Island, a small island about a mile west of Nantucket. The residents of a nearby town, however, were under such "fearful apprehensions" that smallpox would spread from the island among them that they petitioned the General Court to order Gelston to stop the experiments immediately; Gelston had to comply. Similarly, when a group of citizens of Salem and Marblehead established an inoculation hospital on Children's, or Cat, Island off Marblehead in 1773, tensions ran high, and a mob eventually burned the hospital to the ground.[23] The popularity of islands as places to isolate contagion did not subside as the fear of smallpox grew less. In the late nineteenth century, the city of New York used North Brother Island in the East River as a site for a hospital to treat patients with contagious diseases. The facility moved to the island from another insular site, Blackwell's (today Roosevelt) Island. Especially in the case of the latter, the isolation was almost entirely in the eye of the beholder. Far from being remote, Blackwell's Island occupied the middle of the East River, less than four hundred yards away from the shore of Upper East Side Manhattan. The notion that islands are the embodiment of isolation simply because they *are* islands became so well established that at times it substituted for actual physical distance. The popularity of this idea manifested itself in 1858, when the city of New York constructed two artificial islands, Hoffman and Swinburne, off South Beach to house immigration quarantine stations, after dismayed residents of Staten Island burned down the local quarantine hospital. San Francisco started using the natural Angel Island for the same purpose in the 1890s.[24] The use of coastal islands as a simple form of quarantine established them as places whose defining quality was isolation and that, because of this quality, were perceived as well suited for containing disease and separating the sick from the healthy.

The perception of the Boston Harbor Islands as sites that were naturally suited to divide health and disease soon broadened from the literal medical sense to the figurative social one. In the latter half of the nineteenth century, island locations served to articulate social isolation in spatial terms. After South Boston ceased to be as much of an urban fringe as previously, many of the public institutions located there moved on to the harbor islands. Deer Island received the House of Reformation, a prison for juvenile offenders, as well as an asylum for female paupers, and in 1896 these institutions were joined there by the Suffolk County prison.[25] On Rainsford Island, the quarantine hospital was succeeded first by a state- and after 1872 a city-run almshouse; the neighboring and much larger Long Island was used for the same purpose starting in 1885. In 1905 the island was also considered as site of the new state prison. Its dual qualities of relative proximity to the shore and at the same time its separation from it seemed to make it an ideal location for such a purpose. "All the requisites to nearness to outside help seem to exist in the Long island location," declared Police Commissioner Albert T. Whiting, praising the advantages of the island site, "and the natural guard of water will always be on duty and will always be faithful to its trust. The water protection will never fail, and there will be land in plenty for all the agricultural and other needs of the institution."[26] Lack of space elsewhere and the protest of mainland communities that these institutions would likely have drawn, coupled with the convenience of simply maintaining the status quo, made the public institutions on the harbor islands outlive many of the perceptions that put them there. The prison on Deer Island was removed only in the 1990s during the construction of a new and larger sewage-treatment plant, and, in a clear case of path dependency, Long Island remained the site of several public institutions, among them a hospital for the chronically ill and a homeless shelter, until 2014 when the bridge connecting it to the mainland was deemed structurally unsound, closed, and eventually torn down. The site of the Long Island hospital not only inspired the book and film thriller *Shutter Island* but also perpetuated the unquestioned connection between social and spatial isolation deep into our own time.[27]

Dealing with Waste

Keeping diseases, social ills, and the people who carried them away from the city streets was not Boston's only problem that the harbor islands and the water surrounding them helped solve. The growing metropolis was producing rising amounts of waste of all kinds: ashes and street sweepings, bones, rags, broken crockery, decaying food scraps, animal droppings and carcasses, and, perhaps most bothersome, human waste. The waste problem had been present in Boston since the seventeenth century, but on a much smaller scale. The rural town of colonial times not only had a far smaller population and therefore less waste to deal with but could make immediate use of a large part of it. Most of the organic garbage would be fed to domestic animals, and manure and human excrement alike would be used as fertilizer. As the town grew more urban, however, and the farms moved into the country, the transportation costs rose and carting manure out to the countryside was becoming less practicable.[28] As the city's population exploded over the course of the nineteenth century, so, too, did the amount of waste it generated. In 1881 waste collectors employed by the city gathered 147,980 loads of household garbage and ashes, 26,000 loads of offal, 52,020 loads of street sweepings, 8,727 loads of cesspool matter, and many thousands of dead small animals.[29] Put all on one imaginary heap, the collected refuse amounted to approximately ten million cubic feet and would have filled 113 Olympic-size swimming pools. Not only could this tremendous volume of trash hardly be disposed of in the surrounding country, but it also changed in quality, making the recycling of a large part of it relatively difficult. Increasingly, Bostonians settled for another ready means of waste disposal: they threw their garbage into the harbor.

Boston's preferred principle of waste management over the centuries of its existence has been externalization, a trait it has shared with the vast majority of human dealings with materials that are, in one way or another, considered trash.[30] Like other coastal cities, it made use of the fact that it bordered on a natural and seemingly bottomless receptacle. Boston Harbor had a long history as Boston's ultimate sink.[31] Owing to its useful quality of washing out with the tide much of the material that ended up in it, Bostonians viewed harbor dumping as a ready solution to the issue of parting with a wide variety of unwanted substances. This practice, obviously, had its drawbacks: in the half-enclosed water bodies closest to the town's shores, tidal action had only limited effect, and most of the refuse thrown into them stayed there and rotted under Bostonians' very eyes and noses. Dr. Alexander Hamilton, a physician from Annapolis who visited Boston in 1744, described the Town Dock, into which most of the town's sewers emptied and which received more than its fair share of offal and other waste from a nearby marketplace, as "a very stinking puddle."[32] The Mill Pond in the north of the town, which in the 1640s had been cut off from the main body of water by a dam, was by the late eighteenth century widely known as a nuisance "full of putrid fish and dead dogs and cats."[33] Decaying animals populating the town's shores, however unpleasant, were not the last of it, though. Besides throwing solid waste into it, Bostonians also built their sewers so that they led into the harbor and that the sewage, diluted by seawater, would slowly float out of the harbor and, most important, out of sight. It was a principle that worked better in theory than in practice: Boston's natural setting was in this respect anything but favorable to the growth of a large population center. The late-nineteenth-century metropolis struggled with its natural environment that, while having been instrumental in the growth of the eighteenth-century seaport, was largely incompatible with modern urban infrastructure and the strains of a large concentration of people.[34] Not only did extensive land making diminish the slope of sewers, complicating the discharge of sewage into the harbor. Instead of washing away with the outgoing tide, the sewage also often settled on the tidal flats surrounding Boston. While tide flats, large plains of bristling green-brownish mud that give off an unmistakable stench of rotting seaweed, are—from an aesthetic and olfactory perspective—rather unpleasant things to be around under even the best conditions, mudflats saturated with raw sewage were something else entirely. Sewage pollution made them into a stinking nuisance and a health hazard for the city.

Bostonians, throughout the city's history, reacted to this clash between urban needs and Boston's natural setting by altering their city's environment. Just as they built wharves to accommodate shipping and leveled hills to make more space on their peninsula, when confronted with facilities turned dumps like the Town Dock and the Mill Pond, they reached for the easiest solution and filled the offending receptacles with earth and other material at hand, thus abating the nuisance and expanding available space all at the same time.

A closer look at the material history of Boston land making reveals that garbage quite literally shaped the city. The Town Dock and the Mill Pond were early examples of the dual role that waste played in Boston's territorial development. Rendered obsolete by the growing size of seagoing vessels and grossly polluted by the town's refuse, the Town Dock was gradually filled in the 1720s. The Mill Pond disappeared from Boston's cityscape approximately one hundred years later when the tidal mills it powered fell into disuse and the problems it caused became too great to tolerate.[35] The Mill Pond, although large, was a mere puddle compared to the enormous 737-acre Back Bay that, after having been dammed off from the Charles River and crisscrossed by dams and railroad embankments, became a stagnant cesspool polluted by sewage and was gradually filled between 1855 and 1882.[36] So would large areas along the city's shores be filled in the course of the eighteenth, nineteenth, and twentieth centuries, because they had lost their usefulness and turned into nuisances.

Besides being the reason for filling, garbage would often enough be itself used as fill. The Town Dock was filled with the material that made it a nuisance in the first place: cartloads of refuse. Later, when the miasmatic theory of disease discredited building on garbage as unhealthy and put a stop to using this kind of fill, household ashes would be used instead, in addition to gravel and salt mud.[37] Deep into the twentieth century, refuse such as street sweepings and ashes, but also household waste of all kinds, including bottles, jars, wastepaper, and broken glass, were used for filling marshlands in East Boston.[38] Boston was a city built on water, but it was also a city built on trash.

The nineteenth-century city had no longer any use for most of the waste it generated. Miasmatic theory discouraged the practices of dumping garbage in the city and on its shoreline and using it as fill and required a stricter externalization of waste. According to the theory, the stench arising from putrefying organic waste and from sewers was not only an annoyance but also a health hazard that needed to be abated. The city's efforts to dispose of waste as well as to manage trades and institutions responsible for "offensive odors" were characterized by a pronounced tendency to make the city itself a waste-free environment and by an almost complete disregard for what happened with the waste once it was brought beyond the city limits.

The disposal of household waste and offal in late-nineteenth-century Boston was a mixture of older recycling practices and the more recent trend of removing garbage from the city. In 1881 the Board of Health in its annual report criticized the practice of feeding offal and other garbage to domestic animals, since it negatively affected their health and later also the health of humans who consumed their meat, milk, and eggs.[39] Recycling of garbage, while, on the one hand, approved of and lauded as a rational way of dealing with the waste problem that salvaged useful materials, was at the same time considered a dirty, bothersome business that should be banned from the city limits. So, in the 1880s, a yard on A Street in South Boston that collected manure, decayed fruit and vegetables, fish heads, and other organic refuse and turned them into compost drew numerous complaints from its neighbors and was eventually closed by the Board of Health.[40] By that time, most of the available dumping grounds around the city seemed filled. Since garbage was no longer acceptable as land-making material, the city embarked on a method of disposal that was as characteristic of Boston's waste management as it was simple: "It was deemed at this time an excellent expedient to dump the refuse material down the harbor, below the Boston Light, so as to catch the first of the ebb-tide, which would carry it out to sea," stated the Boston Board of Health.[41] Starting in 1884, the refuse would be towed out on scows some ten miles away from the city, and prisoners from Deer Island would shovel it into the water: an arduous and extremely unpleasant job that frequently took up to fourteen hours to complete. In 1886 the city bought several Barney dumping boats that made the task considerably easier: the Barney boats were "self-dumpers," equipped with a bottom compartment that opened up and released the trash into the sea. Boston was not the only harbor city to make use of them: New York leased a whole fleet of the boats from the Barney Dumping Boat Company in the 1880s.[42]

Sewage that polluted tidal flats around the city and poisoned the atmosphere was regarded as still a graver problem of Boston's nascent sanitation. Until the late nineteenth century, Boston's drains and sewers formed a haphazard maze of privately constructed pipes that had only one thing in common: they led down to the city's waterfront and emptied into the shoreline. Constructed to handle storm water and household wastes, they were overwhelmed by the addition of human waste that came with the introduction of water closets in place of privies and outhouses.[43] The municipal sewerage system that had been planned in the 1870s and in use since 1884 differed from the old private sewers in scope but not in principle. The Main Drainage System carried wastewater out of the city toward two outlets in the harbor: the south system brought sewage to a pumping station at Calf Pasture, which conveyed it farther to a reservoir at Moon Island, while the north branch ended in Shirley Gut, a narrow channel that separated Deer Island from the mainland. From both points, sewage was discharged with the outgoing tide.[44]

The metropolitan sewer system not only perpetuated the age-old practice of draining the city's waste into the surrounding waters but more than ever transformed Boston Harbor into an urban facility. It made the harbor's natural tidal cycle into a function of the impressive engineering network that connected the harbor to the city through a system of underwater tunnels and pipes of a perimeter larger than a man's height. The squat Richardsonian Romanesque building of the pumping station stood as an urban outpost on the empty plain of Calf Pasture, a vanguard of technology seeking to include the harbor's shores and waters in the city's infrastructure. Its engines, each with the capacity of pumping twenty-five million gallons of sewage per day, lifted the wastewater thirty-five feet and sent it toward the outfall pipe at Moon Island.[45] A 1875 plan of the future sewerage system shows its two main branches spreading like tentacles around the whole northern half of the harbor, claiming the enormous natural system for the city's needs.[46]

The harbor seemed big enough to receive not only the sewage that flowed through the newly constructed pipes but the contents

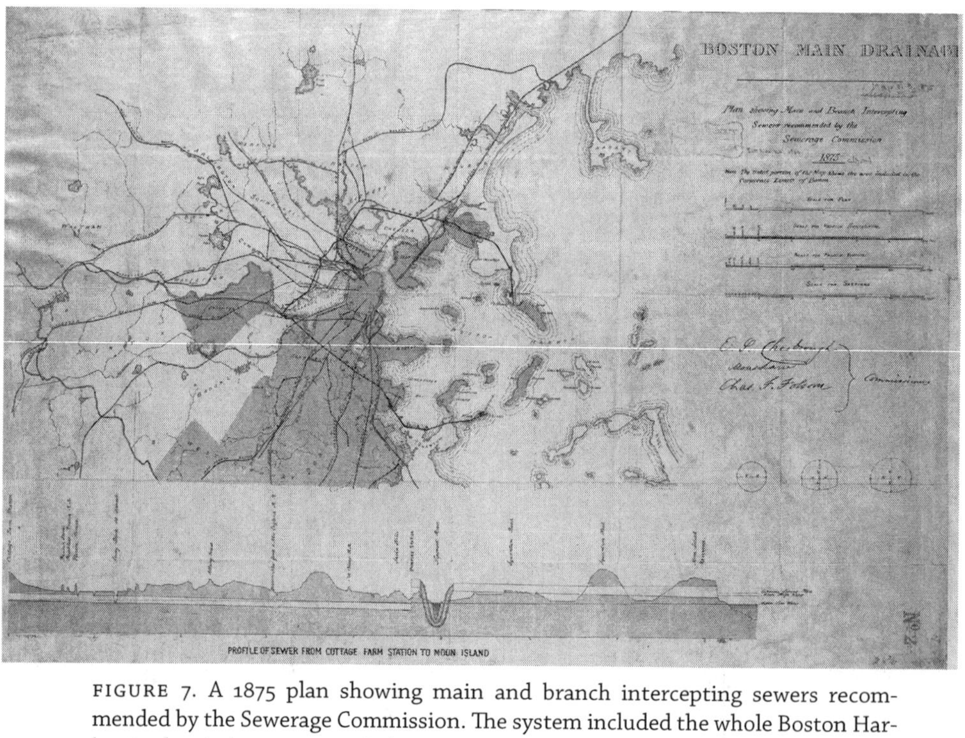

FIGURE 7. A 1875 plan showing main and branch intercepting sewers recommended by the Sewerage Commission. The system included the whole Boston Harbor in the city's wastewater infrastructure.—*Courtesy of Boston City Archives.*

of the citizens' private cesspools and privies as well. In 1884 a private contractor was hired by the city to collect what the contract parties euphemistically referred to as night soil. When the contractor faced allegations that instead of moving the unsavory load out of the city as required, he simply dumped it on the city's mudflats, the Board of Health inquired into the matter. Its findings as to where the night soil went were not surprising: it was transported by scows down the harbor to the deepwater channel off Moon Head, about six miles from Boston's shores, where it was discharged with the ebbing current and carried off to the open sea.[47]

Dead Animals in the City

Late-nineteenth-century Boston was not populated by humans only. Generations after generations of dogs, cats, and rats lived and died in the city's streets, yards, and sewers; their dead bodies were regarded as garbage and were either thrown into the harbor with the rest of household waste or illegally used as fill.[48] Yet beside these familiar urban dwellers, the nineteenth-century metropolis was home to species that have since largely disappeared from cityscapes. In colonial Boston, goats, pigs, cows, and sheep outnumbered human inhabitants several times over; perhaps more surprising, they continued to be present in large numbers even in the urbanized settlement of the late nineteenth century. An inspection conducted by the Boston Health Department in 1892 revealed 2,348 cows living within Boston's city limits.[49] Although the cattle were probably populating mostly the outlying districts, this number is quite remarkable in a modern urban setting.[50] The presence of another animal, however, was still more conspicuous: without the horse, Boston's explosive growth and rapid urbanization would have been unthinkable. Horses were vital for the city's infrastructural and economic functioning: they pulled the streetcars that made Boston's neighboring towns into suburbs, and they ensured the delivery of goods and of municipal services.[51] In 1889 Boston's largest street railway company owned some eight thousand horses, all stabled in the city. The city of Boston itself was one of the largest owners of urban equines: each of the departments of the municipal government had several hundred horses in its employ.[52] In 1880 almost fourteen thousand horses were trotting and trudging through Boston's streets: one horse for every 26.3 people.[53]

When these urban horses died, their massive carcasses not only presented a sanitation problem but were also regarded as source of a number of valuable by-products. Given the sheer number of urban horses, businesses developed that specialized almost entirely on processing the horses' dead bodies. In Boston the chief of them was a company founded in 1828 by Nahum Ward in West Roxbury. With that community's ongoing suburbanization, however, and given the offensive nature of the business, which made it a subject of near-constant scrutiny by the city's Board of Health, Ward was obligated to relocate his plant to a more remote spot. His search for a location that was sufficiently isolated so as not to cause a public outrage yet accessible enough to be practicable for conducting the business had a characteristic outcome: in 1857 Ward bought Spectacle Island from absentee landlord Oliver Whipple, moved his company there, and set up a rendering plant on the west shore of the island's northern drumlin.

By the 1890s, the company was handling two thousand horses a year and developed a stable business routine. Every time a horse died on the street or in one of the city's many stables, the owners or the policeman on the beat who spotted the carcass telephoned the Ward Company. The company owned a large closed cart, and this horse hearse set out immediately to collect the deceased animal and bring it to Ward's wharf on the waterfront. There the horse's mane and tail were cut off and later used for stuffing mattresses and furniture. The now-bald corpse was then loaded onto a boat that went out every evening to Spectacle Island. Once there, the unfortunate animal was in a number of steps completely recycled. The hide was rendered and sold to manufacturers of upholstery, shoes, and baseball covers. After the company's workers removed the animal's skin, the carcass was cut up and cooked in airtight tanks where the tallow was extracted and used in manufacturing candles or soap. The flesh and bone residue were further processed into dog food and fertilizer; the hooves yielded glue.[54] Spectacle became notorious as a "busy colony of manipulators of defunct animals," and dead horses represented only a small part of the organic waste handled on the island. Daily, scows brought twenty to thirty tons of animal waste matter to the island, mostly cattle bones, offal, and suet collected at the city's markets that were turned into tallow and beef scraps.[55] Over time, Spectacle developed into something of a rendering hub for Boston: when a sixty-five-foot dead whale washed ashore near Cohasset, a town on the South Shore twenty miles from Boston, in 1903, it was suggested that the carcass be towed to Spectacle and rendered there.[56]

After the Ward Company moved to Spectacle, within a few decades it was joined there by other businesses that made use of resources extracted from the city's garbage. In the 1880s and 1890s, Ward's tenants on Spectacle included the Dole Fertilizer Company and the LePage Company, a manufacturer of glue and poultry food.[57] The concentration of waste businesses on the island depended partially on the shared resources that they passed on to one another. Increasingly, though, the island became the default solution for locating the city's nuisances. In 1898 a garbage-reduction plant meant to alleviate Boston's problem with the rising amount of waste and the diminishing number of places to put it went into operation at Calf Pasture, a site on the Dorchester shore of Boston Harbor that already hosted the sewage-pumping station.[58] After receiving numerous complaints about offensive odors every summer season from Dorchester's political representation and residents alike, the city moved the plant to Spectacle Island.[59] By 1911 the ensemble of offensive trades on Spectacle grew to include four businesses that occupied the west side of both drumlins and the narrow saddle between them. There was the old Ward Company plant devoted to the processing of dead animals, grease, and bones collected in Boston's streets. There was the garbage-reduction plant that extracted grease from garbage through heating and passed on the dry residue to the somewhat vaguely named Continental Export Company, which subjected it to yet another set of chemical processes for the extraction of more grease and the production of fertilizer. The cluster of waste businesses was rounded off by the International Glue Company, which collected and treated stale fish from which it recovered fish oil and converted the rest into fertilizer and glue. As the city's chief sanitary inspector noted, "These four plants could not be endured on the mainland in the vicinity of habitations or business and can be made tolerable at a distance on the island only by frequent admonitions."[60]

Invisible Garbage

Boston's best efforts were directed toward making garbage and nuisances invisible, making them disappear from the cityscape. If their presence could not be helped at all, then at least they had to be deodorized and neutralized as best as possible. Boston's municipal government shared ideas common to most American cities of the late nineteenth century about the detrimental effects of squalid environments on the bodily and moral health of the population, especially the immigrant one, and about the essential role sanitation played in a well-ordered city. Dealing with waste thus had both an environmental and a social component that made the issue all the more pressing.

The disappearance of the city's garbage happened on multiple levels, from the spatial over the temporal down to the linguistic. In 1893 the Health Department made a three-year contract with the J. A. Rand company of Brockton for the removal of night soil. While the contract specified in great detail that the collection had to proceed in "air-tight and odorless" containers, it did not give any advice about what was to happen with the collected human waste, other than that it had to be removed beyond the city limits.[61] The collection could be performed only under the cover of darkness between ten o'clock in the evening and one hour before sunrise. Night was also the time when most rendering works and tallow- and bone-boiling plants went about their business (rousing the good people of Boston from their slumber in the process).[62] Smelly and unsightly, yet indispensable, businesses such as the city's slaughterhouses would be located outside the city limits, in the case of the municipal slaughterhouse in the then outlying district of Brighton. The efforts to deodorize and conceal the nature of the businesses trickled down to the very level of language. The Brighton establishment was rarely referred to by the all-too-blunt term *slaughterhouse*; instead, the more opaque and elegant-sounding word *abattoir* was used.[63] The smelly yard on South Boston's A Street that was supposed to turn garbage into compost, but mostly had heaps of decayed fruits, vegetables, and putrid offal sitting on the property, giving off a stench fit to wake the dead, was owned by the curiously named Common-Sense Fertilizer Company. The operator of the ill-omened Calf Pasture garbage-reduction plant, an extremely unpleasant business to be near, especially during the summer, was the New England Sanitary Product Company, and it was, perhaps inevitably, the Odorless Excavating Company that was responsible for emptying the city's cesspools.[64]

In cases when garbage could not be ignored completely, it would at least be assigned a clearly defined and ideally remote place. The city's Board of Health maintained a steadily growing catalog of trades not allowed within the city limits: slaughtering cattle and rendering tallow were later joined on the list by manufacturing fertilizer, storing refuse, smoking fish, and refining petroleum, "such trades or employments being nuisances, hurtful to the inhabitants of said city, dangerous to the public health, attended by noisome and injurious odors, and otherwise injurious to the estates of said inhabitants."[65] The nuisances, although deemed indispensable for the city's healthy metabolism, were allowed to exist only at locations removed as far as possible from the residences of Boston's citizens, especially well-to-do ones. While Calf Pasture in Dorchester, five miles from the statehouse and in a district largely populated by immigrants, was deemed a relatively appropriate site for a garbage-reduction plant by the mostly Yankee city government, Spectacle Island seemed an even better place: lacking a constituency and separated from the nearest shore by a stretch of water more than a mile wide, it appeared to be a location perfectly suited for uses that would have created public outrage elsewhere. The Board of Health regulations specified that offensive trades could not be carried out within Boston's city limits, "except on the islands of the harbor."[66] When the state of Massachusetts searched for a convenient location of a sludge- and sewage-treatment plant several decades later, only island sites were under consideration, including Peddocks, Deer, and Long Islands.[67] The islands, sitting as they were in the middle of Boston's ultimate sink, had by the late nineteenth century become the city's officially designated dumping grounds.

In assigning the islands the most bothersome components of the city's functioning—garbage, sewage works, and public institutions for the indigent and the criminal—Boston's city government was trying to translate the desired division between health and disease, virtue and crime, affluence and poverty, and sanitation and filth into spatial terms. In an effort to order and compartmentalize the overwhelmingly chaotic urban life, it strived to create a hierarchy of places in which high and low uses were separated from each other and did not mix. In a system that functioned on the principle of boundaries, dividing lines, and exclusion of certain activities, materials, and people from certain places, the harbor and its islands seemed like places where nature had already conveniently done most of the line-drawing work. The harbor was a marine environment, while Boston was a terrestrial one; the two were separated by the ostensibly absolute division of the shoreline. Even more than Boston Harbor, the harbor islands looked like clearly defined spatial units, isolated from the city by miles of water surface. At least on the map, the harbor and the islands seemed separate from the city proper and therefore ready to receive anything that was not welcome within the city limits. While the Millers River and South Bay clusters of offensive trades grew on the city's very body like ulcers, nuisances pushed down the harbor and onto the harbor islands meant nuisances contained and controlled. Waste that was washed out with the tides or towed out to Spectacle Island was no longer "matter out of place"; it was matter exactly in the place where it was supposed to be.[68]

The Return of the Trash

These geographical hierarchies, however neat they looked in theory, were complicated by the messy realities of the city's environment, society, and infrastructure. Deliberate division of space into separate units that may share a border but do not otherwise overlap was difficult to maintain, or even determine, in Boston's fluid coastal environment. Boston was built in an environment governed by the forces of tides, erosion, and sedimentation, with the boundary between land and sea always uncertain and moving. Human activity made this basic ambiguity even more complex rather than simple: it added land making, dredging, and wharf building to wave action and the impact of river currents, further blurring the boundary between land and water, city and harbor.

Like an unruly corpse, the garbage that should have been safely buried in its watery grave or carried away by the tides kept coming back to Boston's shores.[69] By 1884, the Board of Health came to regard the site off Moon Head as not remote enough for depositing refuse. It sought to remedy matters by simply choosing a dumping site farther from the shore: instead of disposing of garbage in the enclosed waters of Quincy Bay, the dumping was now to proceed near Great Faun Bar off Broad Sound beyond the Boston Harbor headlands.[70] Distance did not help, however: ocean currents carried the floating garbage back toward the land, washing it onto both the North and the South Shores, the coastal regions to the north and south of the Boston Harbor entrance. Moreover, the harbor commissioners and the maritime community in general were complaining that the refuse on the harbor floor was blocking ship channels.[71] The garbage that went out with the ebbing current came back when the tide turned, floating into the inner harbor, sticking around wharves and resting on flats and beaches.[72] When residents of Winthrop on the North Shore complained that Boston's garbage washed out frequently in the town's Crystal Cove, a sheltered bay facing Boston, making bathing "anything but a clean and healthy exercise," Boston's city government dismissed such claims as absurd. The distance between the dumping ground and Winthrop seemed large enough to prevent any unwelcome visits of materials that the city did its best to get rid of.[73] However, several years later it became clear that the complaints were well founded: the harbor pilots discovered that the garbage-scow masters, instead of taking their objectionable cargo to its destination in the open bay beyond the Brewsters, had frequently not bothered to go any farther than barely two miles beyond Winthrop and threw the garbage into the newly dredged Broad Sound Channel. This flew in the face of both the out-of-sight, out-of-mind logic of garbage disposal and the primacy of shipping interests in Boston Harbor. The press was appalled: as the *Boston Daily Globe* titled after the discovery, "Scow Masters Had Used Nice, Clean, Deep Water to Dump Mud and Refuse." Steamships apparently plowed through the soft mass without sustaining any damages, but harbor pilots nevertheless deemed the channel unsafe, refused to take ships through it, and demanded remedies.[74]

The dubious practices of the garbage-scow masters alone, however, could hardly be blamed for the city's ongoing struggle with waste that was clinging obstinately to its body. From the 1910s through the 1930s, the Massachusetts state legislature thought it necessary to establish several special commissions that investigated the discharge of sewage into Boston Harbor.[75] Although the system of releasing sewage into the channel with the outgoing tide seemed clean and easy enough on paper, the tides in the harbor did not function quite as smoothly as when one would flush a water closet. Around the three principal sewage discharge pipes at Deer, Moon, and Nut Islands, sewage was clearly discernible in the water and covered an area of six or seven hundred acres, "while the thin film of oily sleek," one of the sewage commissions observed, "extends to greater distance."[76] Although floating refuse and sewage permeating the harbor water were largely not seen as a serious health risk and harbor dumping was regarded as by far the best and cheapest disposal method available, the practice still created a substantial annoyance to boaters and communities on the harbor's shores.[77]

Environmental boundaries between water and land, however definite and impermeable they seemed on the map, proved much more ambiguous in reality. When Bostonians tried to simplify the complex littoral environment down to a black-and-white division between land and sea, and when they assumed that what they threw into the harbor would stay there, their perception was often thwarted by the harbor currents and the semiaquatic nature of the tidal flats surrounding their city. Similarly, the harbor islands that met the eye as remote and self-contained proved to be far less so when other senses were employed as well. Spectacle Island's spatial isolation that was supposed to keep the annoying side effects of the processing of Boston's waste at bay proved sadly inadequate to the task. The dead horses and later the garbage Boston sent to the island easily traversed the one-mile distance between Spectacle and the nearest shore of South Boston's City Point, although in a different form than they arrived in. Soon after the establishment of Ward's plant in 1857—and for the next one hundred years—Spectacle Island became synonymous with a horrible stench that emanated from the waste-processing businesses and plagued the shores that lay downwind of it. Depending on which way the wind blew, South Boston, Quincy, East Boston, and Winthrop all had to come to terms with the fact that the division between water and land was anything but absolute. Boston's garbage, sent down the harbor to disappear, came back with a vengeance.

The olfactory landscape of the nineteenth-century city was far more various and, frequently, unpleasant than the largely deodorized urban environments of today. Imperfect sewerage and the presence of animals, primarily horses, in the city made the experience of urban life richer in bad odors and the population more likely to tolerate them: a necessity in a world where a large number of people lived close to stables and where horse manure in the streets was an everyday occurrence.[78] The number of complaints made about Spectacle to the city's Board of Health suggests, however, that the odors coming from the island were out of the ordinary even for the hardened nineteenth-century urbanites.[79]

Bad odors had a larger significance for the residents of the nineteenth-century city. As historian Melanie Kiechle shows, the idea that a smell can cause illness was as entrenched in the medical and environmental thought of the day as is today the omnipresent germ theory.[80] Since a stench was understood to be not only the possible signal but also a direct bearer of disease, it was taken far more seriously by the population and experts alike, its presence a health threat and its abatement a medical necessity. Described variously as the "smell of burning animals" and "the odor of food being cooked at high temperature," Spectacle and the odors arising from it were a constant on the city Board of Health's list of nuisances, complained about, monitored, and managed. A Board of Health investigation of Ward's rendering plant in 1873, prompted by a petition presented by City Point residents, revealed a variety of smells coming from open kettles where the horses' flesh was boiling, from fermenting bones lying around on the plant's floor, and from the liquid boiling residue flowing freely over the island's beach.[81] Despite multiple remedies ordered by the board, the plant continued to produce a stench occasionally, especially during the summer. In 1903 Ward's factory was joined at Spectacle by the garbage-reduction plant that was, after only four years, evicted from its original location at Calf Pasture in Dorchester because of the intolerable odors it created. The hopes of alleviating the stench situation on the mainland by relocating the offenders on the island, "where [the plants'] operation will interfere with the health and comfort of no community," did not quite materialize, however.[82] During the following decades, the garbage sent to the island continued to be very much present in the surrounding communities in the form of a nauseating smell that kept returning year after year during the summer season.[83] The smell created a very real—and very disagreeable—bridge across the water that was supposed to separate the community from its garbage.

The association of bad smell with health hazard outlived the miasmatic theory by decades in the public mind. Complaints about "offensive odors" from Spectacle kept reaching the Board of Health, which had in 1891 transformed into the Health Department, well into the 1920s.[84] In the eyes of South Bostonians, East Bostonians, and Winthropites, the emanations from Spectacle, although no longer understood as a source of disease, threatened their quality of life and the desirability of their communities. In their complaints, the meaning of "health" was expanded to refer not only to the bodily well-being of individuals but also to the orderliness of the neighborhoods. J. H. Corney of South Boston, speaking to the *Boston Daily Globe* after a particularly vicious brush with the Spectacle odors in 1907, summed up his neighbors' objections succinctly: "I have known people to be made sick by the smell and it is no exaggeration to say that it is a menace to the people of this very respectable community."[85] He and others who spoke up against the nuisance drew in their minds a clear dividing line between the community of South Boston and Spectacle Island: to them, the one mile of water separating the island and the neighborhood meant also the division between cleanliness and filth, value and worthlessness. While South Boston was a good, respectable place worthy of beautification, Spectacle was a smelly, objectionable, irredeemable waste. In a well-ordered city, the argument seems to have run, every place had its position in the geographical hierarchy, and it was the city's responsibility to control the "bad" places and prevent them from spoiling the lives and the property values of the residents of the better ones. In a rather typical if amusing incident, a person living in Squantum who had one summer morning in 1928 been awakened by "a very offensive odor" from Spectacle called the police headquarters, "asking if the police could not stop the smell. The officer in charge said he was powerless to change the direction of the wind."[86]

The original siting of the Calf Pasture garbage plant had definite elements of environmental injustice: Dorchester and South Boston, the communities nearest to it and most often exposed to the "nuisance," were in the late nineteenth century largely immigrant neighborhoods that were still not as densely built up as the central city. Locating a well-known nuisance in the district could be read as a clear case of the Yankee elite dumping undesirable parts of city infrastructure on people whom they considered inferior and who were underrepresented politically.[87] However, by the time the plant relocated to Spectacle, this situation had somewhat changed: in 1901 when the decision to move the plant was made, the city hall was headed by Patrick Collins, Boston's second-ever Irish-born mayor, who at the time was seeking reelection, relying not least on the South Boston constituency. Spectacle was chosen as the site of the garbage plant because the already existing ensemble of "very offensive plants" seemed to make it a natural place for such uses. The island seemed like the most accommodating and least offensive option: a place so bad that it could hardly be made any worse by the addition of yet another objectionable business.[88]

After germ theory did away with the older connection between disease and foul smell, odors from Spectacle were cast in the role of an amusing annoyance, especially for those who did not have to breathe them in. In the summer of 1922 when the town of Winthrop, some three miles north across the harbor from Spectacle, was experiencing a prolonged period of bad odors from the Spectacle garbage plant, the *Boston Daily Globe* ran a whimsical article about the "gas barrage" that the people of Winthrop had to put up with. "I have been through some pretty tough places in France," the article quoted a "seasoned veteran's" response to a complaining Winthropite, "but that odor from Spectacle has them all stopped. I'd suggest that a requisition be sent to Secretary of War Weeks for gas masks for use during the 'attacks.' They might save some fellows from losing their appetite." The author of the article even offered a poem about Winthrop's olfactory misfortunes:

> While the wind's round West,
> All is well;
> But let that wind from
> South'ard dwell,
> And Winthrop's joys go
> Plumb to hell.[89]

While the grave danger of "some fellows losing their appetite" was a far cry from the notion of whole neighborhoods falling victim to disease transported by the bad odors, and while smells were now just as likely to be the subject of a complaint as of a comic ballad, they were still very much present and noticed in the communities on Boston Harbor's shores. While no longer deadly, but merely unpleasant, the stench lingered and impacted residents' way of life and their communities' reputation.

Whether it came back as a health hazard or as a comic nuisance, Boston's garbage that the city expected to get rid of by dumping it into the harbor or shipping it onto the harbor islands obstinately kept coming back. In spite of Bostonians' best efforts to let their waste disappear entirely, it transgressed seemingly clear environmental boundaries, floated back to the town's shores, and wafted back in the form of intolerable stench, coming to haunt the communities that created it.

Boston's Urban Islands

Despite their ostensible separation from the mainland, the harbor islands' use as Boston's dumping ground drew them functionally closer to the city than they had ever been before. The seventeenth-century town used the islands as resources and as a natural part of its space; the nineteenth-century metropolis integrated them into its infrastructure.

Although arguably not a place most Bostonians wished to visit, by the 1920s Spectacle Island, at least in terms of infrastructure, was an undeniable part of the city. Boats of the garbage company traveled between Spectacle and the mainland several times a day, carrying cargo and providing access for people.[90] The island was connected to the mainland by power and telephone lines, making for the rare sight of telegraph poles lining the island's shores. Its water supply, installed in 1903 at the latest, and probably earlier, came from Boston as well.[91] The other harbor islands fared similarly: Long, Rainsford, Gallops, and Georges Islands had been connected to the municipal water supply already by 1895, with reservoirs constructed on the islands for the use of the public institutions they housed.[92] A small steamer was plying daily between Boston and the quarantine station at Gallops Island.[93] Some of the islands were completely taken over by the demands of the city's functioning: Moon Island, a small island near the Dorchester shore, was permanently attached to the mainland not only by such relatively inconspicuous means as power lines but also by a robust causeway built to carry one of the two large sewer outlets of Boston's newly constructed main drain in the early 1880s.[94] Several years later, four large granite tanks in which sewage was supposed to settle before it would be released into the harbor were built on the island. The tanks, each of them 900 feet long, 150 feet wide, and 17 feet deep, took up half of the island's area and replaced its original dominant feature, a hill overlooking the harbor. "For two and a half centuries, it was the most conspicuous object in this part of the harbor," recalled one contemporary guidebook author, "with its high and grassy bluff rising boldly over Quincy Bay, and dotted with grazing cattle. But now it has become the scape-goat, on which the ills of Boston are to be laid, to be borne off thence into the wilderness of the sea."[95] The island's transformation, while mourned by some, tied it physically and functionally closer to the city.

Boston not only attached some of its harbor islands infrastructurally but also substantially altered their shape and environment. The islands closest to Boston's shore, which until the mid-nineteenth century had mostly retained their former pastoral identity, transformed into landscapes dominated by industrial and infrastructural uses. Moon Island, with the sewage tanks stretching from one end to the other, was one such case; Spectacle Island was another. When Nahum Ward decided to move his rendering plant to the island in 1857, Spectacle, then about forty-two acres in size, was a peaceful, rural landscape. The old stands of trees were now completely gone, having fallen victim to the maritime town's hunger for timber. In the 1830s, the island was inhabited by the Cutler family; its nine members lived in a house on the northern

FIGURE 8. Road on Spectacle Island in the 1930s, with the boardinghouse on the left and the Coleman Brothers' plant on the right. Its use as Boston's waste-processing hub turned Spectacle into an industrial, urbanized landscape. —*Courtesy of Boston City Archives.*

drumlin and farmed the island. Even fifty years later, there was still a large barn standing on the summit, a remnant of the old farmstead.[96] The place's pastoral quality even attracted boating parties from Boston who would come out on the weekends and picnic on the island. This all changed now. The rendering plant altered the olfactory environment of the island, replacing the "healthful sea breezes," one of the island's main former attractions, with the nauseating smells that accompanied the rendering process.[97] The Ward Company occupied a substantial portion of the northern drumlin, having built the factory on its western-facing slope and a small colony of houses for its workers on the eastern one, which soon became known as "Wardville."[98] In the course of the next five decades, it was joined on the island by several other waste-processing companies that clustered on the southern portion of the island, creating an industrial enclave on Spectacle. Each of the companies built its own wharf, so that in 1915 there were five of them extending into the deeper water of the island's west shore. The factories on the island gave rise to a small settlement that housed the workers and their families.

All this activity gave the island a more urban character than it had had ever before. Separated from the city by its unsavory uses, these very uses were responsible for transforming Spectacle's environment from a rural into an industrial and more city-like one. An early-twentieth-century bird's-eye view of Boston Harbor captures the dichotomy of the island's position. It presents the spectator with an environment in which the city is hardly present yet which is decidedly urban in nature. The image shows Boston Harbor and its islands, looking toward the ocean. Only Boston's south shore, in a skewed perspective, is visible in the foreground; the focus is on the harbor islands, which seem to exist independently of the city, taking up the central portion of the picture. They are portrayed in far greater detail than the city whose features are only schematically hinted at. Yet on a closer look, the islands mirror the mainland: Spectacle's factories are carefully rendered, complete with thick black smoke coming out of their chimneys. This is an exact reflection of the way the image presents Boston, as an industrial place dominated by factories, warehouses, and wharves lined by steamships. The other islands as well are all alive with buildings and wharves, suggesting a semiurban landscape. By making use of their remoteness, the city in fact brought the islands closer. It transformed them into working or, in Spectacle's case, even industrial landscapes and made them into reflections of the city's own features.

In the late nineteenth and early twentieth centuries, Boston Harbor was part of the everyday experience for a large number of Bostonians. Although the port's relative position slowly declined compared to other major U.S. seaports, maritime trade, shipbuilding, and other port-related jobs as pilots, longshoremen, and fishermen gave employment to an army of workers.[99] For them, the harbor was a familiar, well-known space that held little mystery. They were acquainted with its topography, its channels and shoals, and they relied on its tides and currents for their living. The harbor

FIGURE 9. This 1923 bird's-eye view of Boston Harbor presents the harbor islands as no different from the city in their landscape and uses. Smokestacks dot several of the islands, indicating their takeover by the city.—*Map reproduction courtesy of the Norman B. Leventhal Map and Education Center at the Boston Public Library.*

islands were part of this experience. Through their location in the middle of Boston's principal working landscape and through their sometimes unpopular yet necessary uses, the islands were places familiar to most Bostonians. Many Bostonians related to the harbor and to the islands through work and perceived them as natural, although in some cases unattractive parts of the city.

In the 1920s and 1930s, the composition of the businesses at Spectacle changed. By the early 1920s, the Ward rendering establishment had ceased operation, leaving the garbage-reduction plant the largest factory on the island. It was operated by the Coleman Brothers' Disposal Company, a contractor that had been in charge of disposing of Boston's garbage since 1922.[100] The Colemans extracted grease from garbage by heating it; the residue was burned and deposited on the northern portion of the island as fill. As the demand for grease diminished and the process became too expensive to turn a profit, the Coleman company went from recycling to dumping raw garbage on the island. In 1935 the grease-extraction plant closed down, and only the dump remained, ever growing.[101] Scows brought the garbage over from Boston, and a crane mounted on a lighter deposited it along the water's edge as fill. The company also built a small railroad that transported the refuse from the wharf at the middle section of the island to the northern summit. A crew of workmen and a tractor were employed for leveling and distributing the garbage.

The landfill operated for only three decades, from the early 1930s until 1959, but it surpassed all previous uses in changing the island's appearance. In a continuation of the longtime tendency of sending waste down the harbor and the last and crudest expression of Boston's disregard for Spectacle as a place that had some value in its own right, Boston transformed the northern and middle sections of the island into an open garbage dump. The island received an estimated 350 tons of refuse daily, and by 1946 its size increased by five acres, made up entirely of garbage.[102] By 1950 most of the structures on the northern drumlin, including the remains of the Ward rendering works, had sunk into refuse. The previously

FIGURE 10. Spectacle Island looking north, probably in the early 1940s. The garbage-reduction plant, which operated between 1903 and 1935, still dominates the island's landscape. By this time, the island functioned as Boston's principal landfill.—*Courtesy of Dolly Snow Bicknell. From the Edward Rowe Snow Collection, Howard Gotlieb Archival Research Center at Boston University Libraries.*

low and narrow sand spit connecting the "spectacles" had been covered with a layer of garbage seventy feet thick; the deep cove on the island's eastern side had been filled with refuse as well.[103] The island's original features slowly disappeared under heaps of household garbage, bricks, scrap metal, broken glass, and ashes. A culmination of the island's almost century-long status as a "nuisance" and the site of "offensive trades," Spectacle now became identical with Boston's principal dump, a place whose sole purpose was to receive Boston's garbage and that by that definition did and could not have any merit of its own.

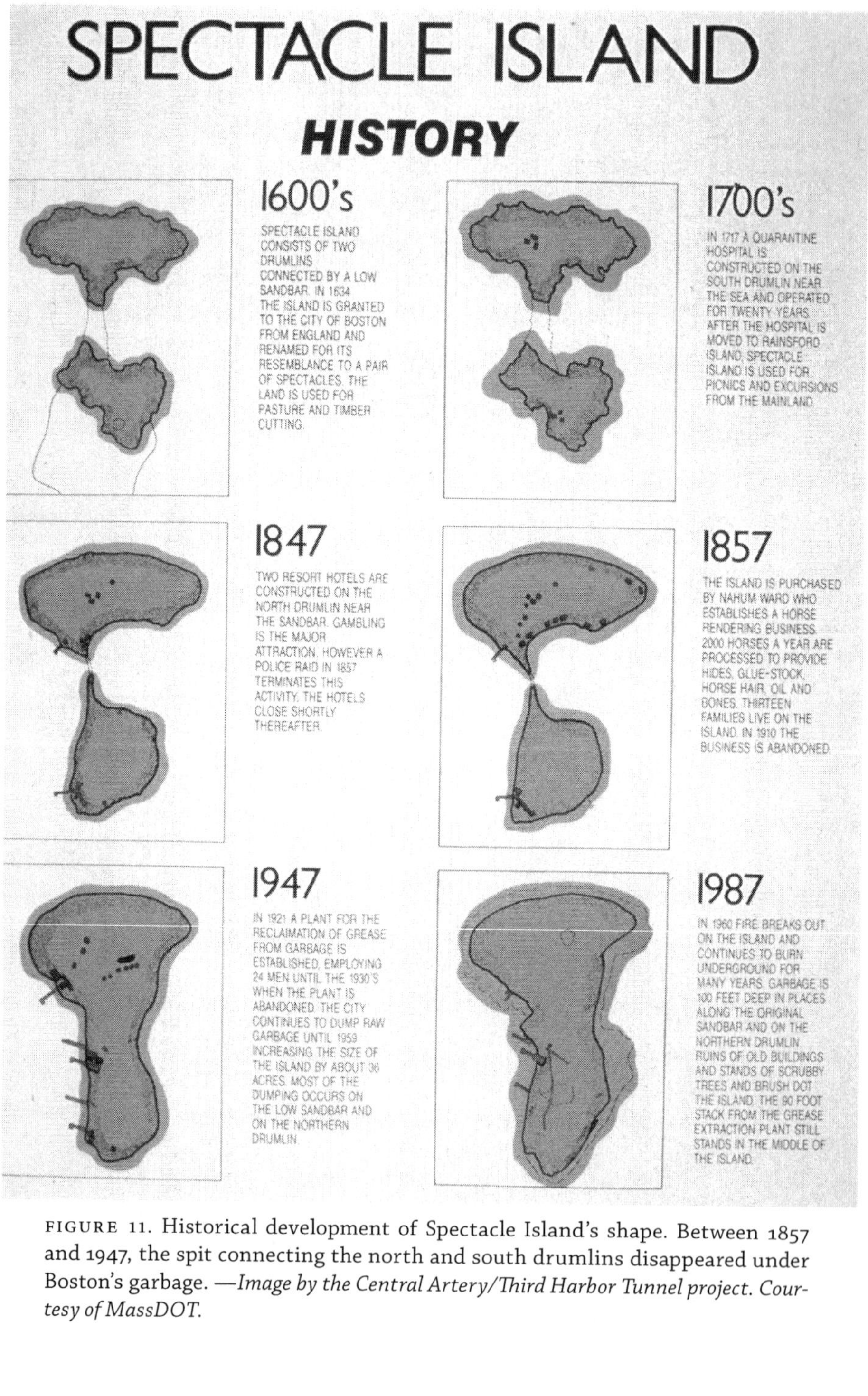

FIGURE 11. Historical development of Spectacle Island's shape. Between 1857 and 1947, the spit connecting the north and south drumlins disappeared under Boston's garbage. —*Image by the Central Artery/Third Harbor Tunnel project. Courtesy of MassDOT.*

The 1948 contract between the city and the Coleman company specified that final disposal of garbage "shall take place only at Spectacle Island," while refuse should be disposed of "by filling of marsh lands, flats, or other suitable locations."[104] The document voiced with astonishing directness the principle quite literally underlying much of the city's development. Wetlands and the harbor islands were at the bottom of Boston's hierarchy of places, remote and inaccessible and therefore well suited to receive the city's trash. Bostonians of the nineteenth and early twentieth centuries, unlike their predecessors, were no longer interested in salt-hay mowing and hunting for water fowl in the wetlands surrounding the city, nor had they any use for timber and pasturage provided by the harbor islands. When the practice of dumping trash onto the flats was criticized at all, it was from the perspective of creating a nuisance, not spoiling a place that might be valuable in any other way. Wetlands lining the shore and many of the harbor islands were treated by the city as vacant land, worthless in their original state and made useful only by providing space for the city's least desirable uses.

Island Life

To those looking from Boston toward the sea, Spectacle Island was nothing but a foul-smelling wasteland. Although it had a history as a pleasant rural site, the rendering works and the waste-processing businesses that followed completely changed its image, and not for the better. As Nathaniel B. Shurtleff, former mayor of Boston and author of the popular *Topographical and Historical Description of Boston*, wryly remarked, "The island, in consequence of the manufactory, has ceased to be so much a place of resort as formerly."[105] To mainland Bostonians, the island seemed a "lonely, wind-swept body of land" that could not be reached "except by traveling over a large space of water" and its inhabitants little more than a curious tribe of reclusive savages whose children "have never seen the streets of this populous city, or looked upon the golden rotundity of the state house dome."[106] By definition of their use as Boston's dumping ground, Spectacle and the other harbor islands could be nothing but an anti-landscape: places damaged beyond repair by human actions and predestined by their unsavory use to be dismal, foul, unlovely, and unfit for human habitation.[107]

Some of the supposed harbor desolations, however, developed quite a happy and self-reliant life of their own. The quarantine station on Gallops Island still admitted an occasional case of yellow fever and detained passengers from ships with "sickness of questionable nature" on board in the late nineteenth century.[108] Other than that, little on the island reminded a random visitor that they found themselves on Boston's pest island. In 1876 port physician Alonzo Wallace reported that the low number of patients in the previous three years presented the staff of the quarantine with a good opportunity to renovate the structures and cultivate the land on the island. By 1877 the hospital had new buildings, all "very neat and substantial," the island's banks were graded, and the beach had been cleared of debris. Even the graveyard of disease victims on the island was brought to order. Each of the 206 graves had been marked with "neat and appropriate head-boards, with inscriptions giving the name, age, and date of death, and the grounds have been enclosed by a good fence," thus making even the grim reality of death conform to the general orderliness of the island. The staff also turned to gardening and husbandry, keeping a small cow herd that supplied the island with milk and butter. Almost all available land on the island was occupied by vegetable gardens and grazing animals. In 1876 the "quarantine island" not only admitted twenty-five patients to the hospital but also produced ten tons of hay, one hundred heads of cabbage, ten bushels of turnips, and "very creditable quantities of potatoes, onions, beets, grain, etc. etc., which are used upon the island and quarantine steamer."[109] The place that by all measures was supposed to be synonymous with disease, infection, and death was in fact a well-ordered, self-sufficient community teeming with life.

Like the pest island, Boston's trash heap was inhabited as well. In 1882 the Ward factory employed 30 men: some of them commuted daily from the mainland, while others lived with their families directly on Spectacle Island. The settlement included some twelve dwellings, a water reservoir on which the residents skated in winter, and most conspicuously a school that by the 1890s had about 20 pupils and was occasionally the subject of curiosity-seeking newspaper articles as "the most picturesque of all the schools in Boston."[110] As the number and size of the factories on the island grew, so did the community. In 1895 Spectacle had 94 residents, and over the next three decades this number grew to 119.[111] By the 1920s and 1930s, a close-knit community of islanders evolved, almost all of them employees of the Coleman Brothers garbage plant. Most Spectacle islanders were working-class people; later, they were joined by middle-class families fallen on hard times during the Depression. The families occupied a handful of scattered houses on the southern slope, while single men employed by the Coleman company lived in a boardinghouse near the factory. Many of them were first-generation immigrants from Poland, Lithuania, Ireland, or Russia. Although "not a fancy place," as one former resident put it, in respect to its social and ethnic makeup, Spectacle did not differ much from South Boston across the water.[112]

In the memories of the inhabitants, the garbage island existed as a surprisingly pleasant place. "I don't remember it smelling

FIGURE 12. House at Spectacle Island looking north, probably in the 1930s. For Spectacle's inhabitants, life on the island was not as dismal as it may have seemed from the mainland. —*Courtesy of Boston City Archives.*

badly," says Kathleen Rivera, who spent her childhood holidays on the island. Unlike the mainlanders, who associated Spectacle with noxious odors, garbage, and the nuisance that made their lives miserable and their property values plummet, Spectacle islanders themselves hardly ever mentioned the olfactory part of their lives. The less pleasant aspects of island life were, it seems, either ignored or turned into something useful. In the 1930s, as the landfill on the northern portion of the island began to grow, the islanders developed quite a steep topographical hierarchy of their own. While the settlement concentrated on the southern drumlin, the residents increasingly avoided the northern part where Boston's garbage piled up. "Spectacle was where we lived," recalled a former resident, Jenny Lescovitz. "The other side of the island we didn't even call Spectacle. We called it Ward's."[113] Sometimes, though, even Ward's proved useful. Robert Wyatt, who grew up at Spectacle in the 1920s and 1930s as the son of Colemans' superintendent Roy Wyatt, remembered how, at times, the landfill turned into a resource as the refuse deposited there washed up on the island's shores:

> A [northeastern] storm was best which brought to the surface many things of value for easy picking such as gold and silver jewelry, also forks, knives, and spoons made of sterling. Of course this attracted us. We became dump pickers. Mother spent much time doing this as did us kids. Junk (all kinds) brass, copper, lead etc. We cleaned this, sorted it all out and fill [*sic*] many containers which we later sold to a Boston junk dealer.[114]

Living on a harbor island offered opportunities that life in the city's congested, parched streets did not. Former residents uniformly remember Spectacle as a green, pleasant place that profited from its harbor setting and the fresh sea breezes and compared favorably to the crowded city on the mainland. Most families on the island cultivated vegetable and flower gardens and also owned boats that allowed them not only to access the mainland after the Coleman company boat stopped running at 4:30 in the afternoon but also to travel around the harbor, to fish for lobsters and flounder, or simply to go on "vacation" in a different corner of the harbor.[115] Even when taken with the proverbial grain of salt, the memories of the islanders suggest that life on Spectacle looked very different from the inside than it did from the outside. The experience of the inhabitants of the supposed anti-landscape was in sharp contrast to the dismal image the island had acquired for those who looked at it—and smelled it—from the mainland. Deeply attached to their peculiar home, Spectacle islanders were, decades after they had left the island, anxious to communicate their memory of Spectacle as more than only a smelly trash heap. Several of them wrote down their recollections of life on the island. Robert Wyatt, who spent the first twenty years of his life on Spectacle before moving to Boston with his wife in 1942, accompanied his memoirs by an elaborate map of the island that detailed the location of every house, road, mussel bed, favorite swimming spot, and boat mooring place. The Wyatt map is easily the most detailed and accurate rendering of the island as it looked in the 1930s—and it differs dramatically from the few plans of Spectacle that the city of Boston had on record. While the city's plans remained strangely vague and sketchy, Robert Wyatt's map of Spectacle covered every square inch with details about the topography of the island and the adjacent waters. Whereas the municipal plans gave only the location of wharves and factory buildings, Wyatt provided not only the exact location of every house and the names of its inhabitants but also a sense of place defined by human activities and natural conditions: we learn that there were two good spots for swimming at the southern tip of the island, that one could fish from the company wharves, and that the northeast wind habitually brought rain or snow. While the city made Spectacle look like nothing more than a site serving a specific function, a facility devoted only to the purpose of processing and depositing of garbage, Robert Wyatt's map presents the island as a storied landscape, made unique through relatable human experience.

By the early 1940s, when all the factories had closed and only the landfill remained, Spectacle's permanent population dropped to about ten people. The island, while during the past century never

FIGURE 13. The Timmons family in their garden at Spectacle Island, 1934. Most families living at Spectacle cultivated their own vegetable and flower gardens. —*Courtesy of the University Archives and Special Collections Department, Joseph P. Healey Library, University of Massachusetts, Boston: Boston Harbor Islands Mass. Memories Road Show Collection.*

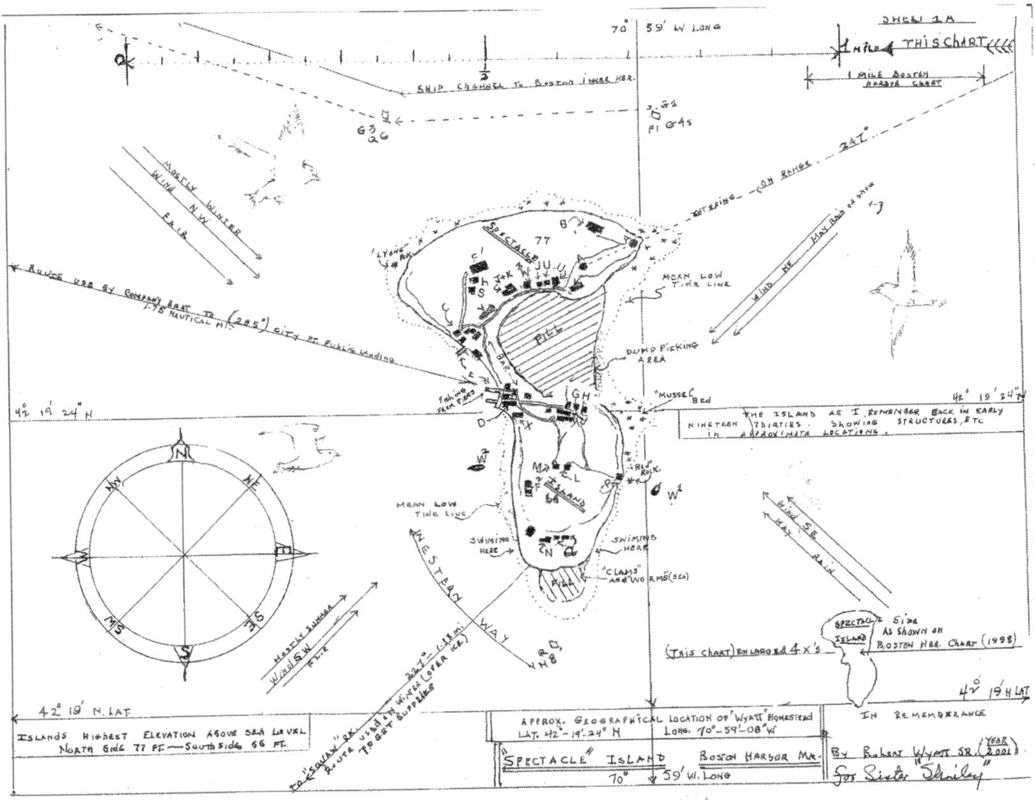

FIGURE 14. The map drawn by Robert Wyatt, a onetime islander, shows Spectacle as he remembered it looking in the 1930s. It presents the island as a lived-in place defined by human experience. —*Map by Robert Wyatt Sr.*

quite the most popular spot in the Boston area, was fast slipping into oblivion.[116] The city of Boston, while happy to call on the island's services as a dump for its refuse, was otherwise almost completely ignorant of its existence. Also, the city failed to grasp the extent of the changes its actions imposed on Spectacle Island. In 1941 a U.S. Navy request to use Spectacle for defense purposes led to a frantic search at Boston City Hall for documentation clearing the question of who owned the island. After a flurry of communications between the Law Department, the Public Works Department, and the Office of the Mayor, the city seemed somewhat surprised to find that it owned six acres of land on the island. When the ownership question was cleared, the next puzzle presented itself. Now the officials wondered how big the island in fact was. In the end, the Law Department produced a map from 1915, showing approximate locations of the various waste-processing plants and property boundaries on the island. As this seemed the best available information, the city went with it, completely disregarding the fact that the map was rendered obsolete by the intermittent decades of dumping raw garbage and residue from the factories on the island's northern and middle portions and that the majority of the buildings the plan showed were in ruins by the time of the inquiry.[117] In the same way that atlases continue to show the full extent of a dried-up lake long after it is gone, Boston's municipal administration clung to a decades-old representation of the island, surmising a geographical stability where there was none. Operating under the premise of fixed and unchanging boundaries, the city ignored transformations that were

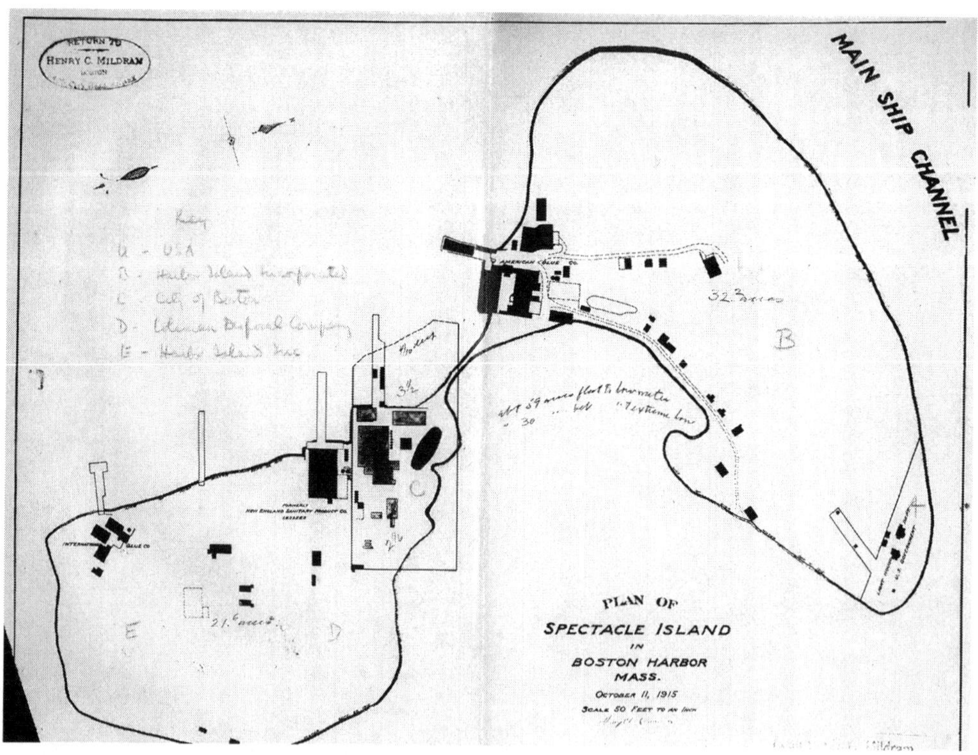

FIGURE 15. A 1915 plan of Spectacle Island had already been rendered obsolete by man-made changes twenty-five years later.—*Courtesy of Boston City Archives.*

of its own making. For city hall, Spectacle Island was a place meant to fulfill a certain function; beyond that, the city cared very little.

Even Spectacle's function as the city dump was fast becoming obsolete. Public health experts had been advocating incineration as the most sanitary method of waste disposal for decades, and with the growing heaps of trash that Boston's population produced and the dwindling number of places to put them, the city was ready to follow through with their suggestions.[118] In comparison to incineration, which promised to solve the city's waste problems in an elegant and sanitary way, leaving nothing but clean ashes, the dump on Spectacle was messy, smelly, and decidedly unmodern. When the city's Financial Commission reported to Mayor Maurice J. Tobin on the matter of waste disposal in 1941, it took a firm stand against continuing the practice of dumping garbage on Spectacle. The operation, the report claimed, was not only unnecessarily expensive but also dangerous to health, unsightly, and foul smelling. Incineration, on the other hand, was "the most scientific and economic method" that would rid the city of the Spectacle nuisance, save money, and even generate heat, all in a perfectly sanitary and unobtrusive way.[119] In 1944 George G. Hyland, the public works commissioner, announced that he was putting the construction of a garbage incinerator for Boston on the city's "must list" to replace the dump on Spectacle Island and called the habit of dumping garbage into waterways and on rubbish heaps at the city's outskirts "a sign of untidy youth." "It is high time," he stated, "to leave off such primitive practice."[120]

In 1959 the long-awaited waste incinerator went into operation in South Bay, another location with a long and rich history of rubbish dumping. The landfill at Spectacle was abandoned.[121] Roy Wyatt, the last superintendent of the garbage company, and his wife left the island as its last permanent residents. After centuries of habitation and active use, and after decades as Boston's unpopular yet indispensable dumping ground, Spectacle Island became vacant land, a place that the city seemed to have no use for any longer.

3

THE CITY OF TOMORROW

Boston Harbor and Visions of Urban Renewal

In late June 1969, Bostonians, after years of neglect, saw their harbor again. They did not see it from the deck of a ship cruising its waters or from a beach on its shore; they saw it from a helicopter. They did not even see it firsthand; they saw it on the screen, in a color film documentary. But they finally saw it—and they did not like what they saw.

The seventeen-minute film *Boston: Harbor/City/Islands* originated as a presentation to the Massachusetts state legislature. In 1969 the legislators reached out to Stanley Jacks, professor at MIT's Sloan School of Management, to undertake a study of Boston Harbor's present state and possible uses. Jacks in turn approached the well-known filmmaker Derek Lamb, who taught film animation at Harvard, and his student Lawrence Rosenblum. Instead of a customary slide presentation, the three of them opted for a more dynamic and immediate way to acquaint the lawmakers with their capital's harbor. In their short film, they seemed to have taken pains to show the deteriorated state that the harbor and its islands were in. They did not have to stray far from Boston's shore to do that: after giving the viewers a brief look at the run-down port district in South Boston and scattered industrial uses along the harbor's south shore, the film crew swung around the relatively idyllic-looking seaside community at Squantum and headed out on the open water. Turning to three of the islands closest to the city, Moon, Long, and Spectacle, they found disuse, degradation, and decay. The sewage tanks at Moon Island provided a visual for the troubling topic of the harbor's chronic pollution, and the crumbling remnants of Fort Strong at the northern tip of Long Island bore witness to the demise of most of the previous uses of the islands and the neglect that followed them. But it was only when the camera swept across the water to Spectacle Island that the desolate state of the harbor islands came across at its most vivid. As the helicopter circled over the island, its whole surface sparkled in the afternoon sun. The glitter was not clamshells or pebbles: it was broken glass covering the site of the former landfill in a solid layer. Ruins of wharves, the hollow and decaying shells of factory buildings, and smoke rising from underground fires completed the picture of an abandoned, forbidding, desolate place. "Much work will have to be done to islands like Moon, Long, and Spectacle, as well as to other islands of the harbor and the waters of the harbor itself before they will be suitable for public use," the narrator concluded in a sepulchral tone.[1]

FIGURE 16. The glittering surface of Spectacle Island in the 1969 film *Boston: Harbor/City/Islands* by Derek Lamb and Lawrence Rosenblum. By this time, the island had been abandoned for ten years. —*Courtesy of Lawrence Rosenblum and the University Archives and Special Collections Department, Joseph P. Healey Library, University of Massachusetts, Boston.*

Regarding the harbor islands as outcasts of the city's geography was a long tradition in Boston. Since at least the eighteenth century, many of the islands had become synonymous with quarantine stations, prisons, rendering works, landfills, and other unsavory uses. To mainland Bostonians, they were the very definition of marginal places, bleak and uninviting.[2] The way they were presented in the 1969 documentary was entirely in keeping with this perception. What was new was how vehemently this situation was criticized in the film. For generations of Bostonians, the harbor islands' function as the city's universal dumping ground was a necessary evil, and they viewed the islands as ideal locations for such uses. The documentary presented a radically different view. To its creators, the fringe uses of many of the islands were "distasteful" and the general way the city treated its harbor and islands nothing less than "misuse and neglect." The harbor, in their view, was in fact "slowly dying." These were harsh words indeed—and a sharp departure from the out-of-sight, out-of-mind mentality that had governed much of Boston's historical dealings with its marine front yard.

Still more remarkable was the notion that the islands had the potential to be redeemed and once again become useful to the city. Despite the partly unattractive visuals, the film presented the harbor and the islands, even in their ostensibly deteriorated state, as comparing favorably to the city in some respects. While the Boston cityscape, as the documentary presented it, seemed made out of concrete, smoking factories, and dreary expressway junctions, the harbor, despite its decried "misuse and neglect," possessed a strange tranquillity. While congestion and blight had hold of the central city, there was both blue water and open space in the harbor. Even the film's soundtrack was employed to help draw a sharp dividing line between the city and harbor environment. Whereas the city produced a cacophony of honking cars and traffic noise, and a medley of radio broadcasts, all of them finally overshadowed by the deafening roar of airplanes, the harbor remained soothingly quiet, its lack of bustle underlined by seagull calls and the transcendent tones of a Bach violin concerto. The message was clear: not only was the harbor essentially different from the city, but it had something the city lacked—and needed. It offered space that could alleviate the city's congestion, and it had a calm that contrasted with the city's incessant noise. "If properly managed," the film's narrator assured viewers, "the harbor and its islands hold great potential." By putting the harbor and its islands to "proper" use, the logic went, the city could enhance its own space, livability, and attractiveness. There were, however, a wide variety of opinions about what constituted proper use. After decades of having been largely absent from the public eye, in the late 1960s the Boston Harbor Islands took center stage in the city's deliberations about its future. This time, however, they were no longer Boston's ultimate sink, pushed mentally as far as possible from the city's waterfront. This time, they were about to become Boston's new marine neighborhood.

The Decline of Boston's Waterfront

The disuse of the harbor islands in the 1950s and 1960s, presented with such grim emphasis in the documentary, was more perceived than actual. The only large-scale decline in the islands' use was caused by the U.S. Army abandoning the forts it had maintained on several of the islands. Using the harbor islands for military purposes was a practice almost as old as Boston itself: the first fortification on Castle Island dated back to 1634. A succession of increasingly elaborate forts was built on the site, which remained Boston's sole harbor defense well into the nineteenth century. By the time the Civil War broke out, Castle Island's Fort Independence was joined in its task of guarding one of the Union's vital ports by Fort Winthrop on Governors Island and, even more prominently, by Fort Warren on Georges Island at the harbor entrance. It was, however, the close of the nineteenth century that saw a boom of defense building activity in Boston Harbor. Following the final report of the so-called Endicott Board in 1886 that called for a comprehensive defense scheme for America's coastal cities, gun batteries were placed on Long, Georges, Lovells, and Peddocks Islands.

These island fortifications were flanked by gun emplacements on the harbor's headlands in Winthrop and at Point Allerton, turning Boston Harbor into a heavily militarized landscape. The War Department also appropriated all the islands it used, becoming one of the largest landowners in the harbor.

Like the city of Boston, the military held an exclusively utilitarian view of the harbor islands, albeit using a different perspective. Whereas the city appreciated the islands' relative remoteness that lent itself to unpopular uses, the War Department saw them as useful outposts that allowed for creating effective fields of fire to meet incoming enemy ships. Once the army appropriated the islands, it was also conveniently free to treat them as it pleased: pave them over with reinforced concrete, dig trenches and mortar pits in their upland, and mount heavy weaponry on them. This approach turned whole islands into one-purpose military facilities. On the eve of the Second World War, new defense structures built on Great Brewster and Outer Brewster Islands included a complex mining casemate, barracks, quarters and messes, a reservoir, a searchlight, platforms for antitorpedo batteries, and a desalination plant. Within three years after the war's end, however, it became obvious that the harbor fortifications had been rendered obsolete by advances in air warfare and by the advent of ballistic missiles. The Boston Harbor defenses, including Fort Strong at Long, Fort Andrews at Peddocks, Fort Standish at Lovells, and the Civil War–era Fort Warren at Georges Island, were decommissioned and the islands either transferred to the state of Massachusetts or sold at public auctions.[3]

Apart from the military's pullout, the scattered institutional uses of the harbor islands remained much the same as in previous decades. The landfill at Spectacle had closed and the buildings of the quarantine station at Gallops and the paupers' house at Rainsford Island stood empty, but the prison at Deer, the hospital at Long, and the boys' farm school at Thompson Island continued operation, as did the centuries-old lighthouse Boston Light at Little Brewster. The tanks at Moon Island still regularly filled with sewage, and they had been joined in the 1940s by a new sewage treatment plant at

Nut Island, a former island connected by fill to Hough's Neck in Quincy. Most of the traditional "unsavory" uses of the islands, in other words, went about their business as usual. The major difference was that most Bostonians were no longer there to see them.

In the first half of the twentieth century, Boston Harbor changed from the city's front door into its backyard. Mostly, national and regional economy as well as new developments in transportation were to blame. Boston, once the nation's busiest port, was in the early twentieth century rapidly losing trade to its East Coast competitors Baltimore, Philadelphia, and above all New York. Boston's longstanding problem, the lack of a bulk export product, was combined in the latter half of the nineteenth century with an unfavorable railroad differential. Since among the major East Coast seaports Boston was the farthest from the production centers of the Midwest, railroad companies required extra freight rates for transportation of goods to Boston, sending manufacturers into the arms of ports that were closer and therefore cheaper.[4] Moreover, in 1916 the North Atlantic Conference of steamship lines introduced equal ocean rates to all North Atlantic ports, effectively doing away with Boston's advantage of being more than three hundred miles closer to Europe than Philadelphia and almost five hundred miles closer than Baltimore. The port of New York, by the turn of the century already a giant dwarfing its competitors, offered more and better services to its shipping, causing even New England products to move through New York instead of Boston by the 1920s.[5] Finally, Boston was hurt by the shifting composition of international trade, as trade in agricultural produce, which made up much of the port's profit in the late nineteenth century, declined in the first years of the twentieth.[6] When combined, these developments spelled doom for Boston's prominent position among Northwest Atlantic ports. Whereas in 1901 Boston occupied second place among the nation's ports in foreign exports, it dropped to sixth in 1920 and to the appalling eighteenth place in 1929. The Great Depression then swept away whatever remained of the city's once-proud maritime commerce, sealing the fate of the port's—and the harbor's—importance for Boston's economy.[7]

The port's economic decline was preceded by its gradual demise as a place where all social classes mixed and that was part of almost every Bostonian's experience. Even in the port's heyday in the mid-nineteenth century, the waterfront became increasingly alienated socially from the rest of the city. As John R. Gillis, an eminent historian of the seacoast, argues, in the course of the nineteenth century waterfronts of large port cities everywhere were becoming places of ill repute that were frequented less and less by "respectable" people.[8] Gone were the days when the stately Long Wharf presented the representative avenue into the city. Now Boston's wharves became realms of foreigners, immigrants, the itinerant labor force of the sea, and illicit activities like gambling or smuggling. While working-class and immigrant Boston was still actively involved with the city's harbor, the middle and upper classes had turned their backs on it.

The effect the port's diminishing importance for Boston's economic life had on the activity in the harbor was augmented by the advent of containerization in the 1960s. Although the port of Boston was at first reluctant to adapt to the new technology, its increasing use was beginning to eat away at the traditional longshoreman workforce.[9] Coupled with automation, these developments reduced drastically the number of people for whom Boston Harbor was a familiar workplace and who engaged with it on a daily basis.[10] The onset of the jet age in the 1960s eliminated even a spectator's contact with the harbor and its islands. With Logan Airport expanding explosively from the 1940s through the 1960s, both in area and in passenger numbers, visitors coming to Boston were now more likely to land on one of Logan's runways than to approach the city on board an oceangoing vessel.[11] In a now bygone era, ocean liners would wind their way among the harbor's many islands at relatively low speed, offering their passengers views of the harbor's varied landscape. In contrast, airplanes would now whizz over the harbor in a matter of minutes, giving the travelers barely a chance to glimpse the water. Thus, within only a few years, Boston Harbor and its islands went from an oceanic gateway to flyover country.

The economic downturn manifested itself physically on Boston's waterfront: too large for the diminished amount of shipping, the district became characterized by crumbling piers, moribund facilities, and empty warehouses.[12] Perhaps even more damaging for the city's contact with its harbor was the construction of the Central Artery in the 1950s. An elevated freeway built through Boston's downtown to improve the traffic situation and to alleviate inner-city "blight," much in the vein of the colossal projects pursued by New York's controversial master planner Robert Moses, the Central Artery cut off the waterfront district from the rest of the city.[13] Running along much of the city's harbor front and isolating the central waterfront and the traditionally maritime neighborhood of the North End, it prevented Bostonians from engaging with or indeed even seeing their harbor.

Boston's Urban Renewal

The Central Artery became part of the problem, but it had been built as part of a solution. In the 1950s, the city of Boston was in dire straits financially. The port was in a decades-long downturn that included the closing of two large shipyards in East and South Boston.[14] Traditional industries, primarily textile and shoe manufacturing, were leaving New England for the lower wages of the South. Even the number of New England fishermen decreased by one-third between 1950 and 1963 due to imports of frozen fish from abroad.[15] Boston's moribund finances could also hardly count on support from the state or investments from Boston capitalists. The city's Brahmin elite that held sway at the statehouse and the largely Irish-run city hall had waged a protracted feud against each other ever since John F. Kennedy's grandfather John F. "Honey Fitz" Fitzgerald was elected mayor in 1906 and were less than inclined to help each other out financially. The central city's situation was not made better by the massive suburbanization that caused the metropolitan region to swell from a perimeter of ten miles from the city hall in 1940 to twenty-five miles in 1980.[16] To no small extent, the infamous practice of "redlining" was to blame for this sprawl. Enabled by the National Housing Act of 1934, local banks were in the 1940s and 1950s classifying many of the city's old ethnic neighborhoods as blighted areas and refused to grant mortgages in them.[17] This practice had all the trappings of a self-fulfilling prophecy: since the chances of getting a mortgage in these districts were extremely low, they soon became just as decayed as the banks had predicted. Prompted by this development and facilitated by postwar federal subsidies on the construction of highways, Boston's white middle-class population relocated to the new suburbs in large numbers. Whereas in 1950 the central city had a population of slightly more than eight hundred thousand, by 1960 this number had dropped to seven hundred thousand.[18] These middle-income residents took their tax revenue with them, leaving Boston struggling to keep its finances afloat and to retain its ability to provide municipal services to its increasingly aging and low-income inhabitants, many of whom had a migration or minority background.[19] The exodus of a hundred thousand people, accompanied by the departure of their jobs and stores that catered to their needs, pushed the city to the brink of bankruptcy by 1959.[20] The erstwhile hub of the universe was reduced to a "hopeless backwater, a tumbled-down has-been among cities."[21]

Boston understandably tried to push back. It chose to do so by the means most readily available to cities all across the postwar United States: by banking on the opportunity offered by the 1949 and 1954 U.S. Housing Acts. These laws allocated federal funds to projects aimed at rehabilitating "slums," reducing "blight," and making American cities attractive environments for business and well-to-do white residents.[22] Shortly after the election of John B. Hynes as mayor of Boston in 1950, the city embarked on a wholesale urban-renewal program. Hynes, who succeeded James Curley, a colorful Irish American political figure renowned for his unorthodox political tactics and the delight he took in skirmishes with the Yankees at the statehouse, was able to win the support of the state government and the city's economic elite and secure their cooperation with the municipal administration.[23] In an address in 1954,

Hynes laid out a plan for a transformed "New Boston": he called for slum clearance, a redeveloped downtown business district, improved public housing and recreational facilities, and revitalized neighborhoods.[24] During the 1950s, the city developed an ambitious urban-renewal program that was designed to attract middle- and upper-income residents and investors. In 1957 Mayor Hynes in cooperation with the state legislature established the Boston Redevelopment Authority (BRA), an agency that was to supervise the city's makeover. When John F. Collins took office as Boston's mayor in 1960, he followed in the footsteps of his predecessor. He almost immediately brought to Boston Edward J. "Ed" Logue, the energetic, trailblazing urban planner who had until then been in charge of the urban-renewal program in New Haven, as the new head of the BRA. Logue lost no time: in September 1960, he and the mayor unveiled the "Ninety Million Dollar Development Program for Boston," a largely federally funded comprehensive redevelopment plan that proposed to redesign a staggering 25 percent of the city's acreage.[25] Over the next decade, Boston found itself in the middle of a building frenzy. The old town that in the eyes of municipal officials seemed to consist of tenements, out-of-use commercial structures, and somewhat seedy districts was giving way to a modern metropolis of steel, glass, and, supposedly, prosperity. The rationale for this dramatic transformation stayed much the same throughout the 1950s and 1960s. The BRA's *General Plan for the City of Boston*, published in 1965, listed as the city's major troubles the same familiar culprits: "economic stagnation; antiquated public services; an inadequate transportation system; loss of population; deteriorating housing; and extensive commercial and industrial blight."[26] By then, backed up by federal funding, Boston's urban-renewal program was progressing swiftly, seeking to reverse the city's fortunes by changing its physical and social makeup.[27]

The most readily discernible face of Boston's urban renewal was the office towers mushrooming in the city's central districts. Prudential Center in Back Bay, the high rises of Charles River Park, the State Street bank that started the movement of financial institutions into Boston's downtown, or the I. M. Pei–designed Harbor Towers were all monuments to the city's effort to keep its tax base, in which private business played no small role, sound and growing. The city administration did not forget about itself, either: a new government center went up in the very heart of the city, its centerpiece the imposing structure of the new city hall. Although ridiculed by its critics as "the crate that Faneuil Hall came in," the brutalist building was an apt expression of the magnitude and the uncompromising determination with which the city pursued its renewal plans.[28]

There was, however, also a far side to Boston's urban renewal that was less monumental but significantly more problematic. Referred to euphemistically as "slum clearance" or "blight elimination," whole neighborhoods were razed to the ground to make way for the new commercial and institutional development. First to go were the so-called New York Streets, several city blocks in the South End district named after towns in upstate New York to commemorate the opening of the Boston–Albany railroad in 1842. Between 1955 and 1957, the multiethnic neighborhood was demolished and converted to industrial use. However dramatic, the extent and impact of the New York Streets' demolition paled in comparison with the West End renewal project that succeeded it. One of Boston's oldest immigrant communities, the West End was home to approximately twenty-three thousand people of Irish, Italian, Greek, Armenian, Polish, and Russian Jewish descent. Depending on the point of view, its narrow, crooked lanes were either Old World picturesque or dilapidated, overcrowded, and filthy. In the eyes of the young urban planners, they were the latter: in 1958 the BRA launched the West End's "redevelopment." Before the community had even time to mount a substantial protest, the bulldozers had already moved in. The district's tenants were displaced without adequate alternative housing provided for them. Within less than two years, most of the neighborhood's approximately nine hundred structures and twenty-seven hundred households ceased to exist, and the area was redeveloped as a middle- and upper-income enclave.[29]

The demolition of the West End may have seemed a "dilapidation to delight" story to its architects, but it was in the first place a human tragedy and also a public relations disaster. The Boston public was horrified by the city's ruthlessness during the evictions and the demolition, and the waves of outrage that the West End's fate created still rippled through other renewal projects years later. While there was no resurrecting the old West End, its brutal demise spelled an end to drastic, large-scale redevelopment plans without community involvement. It boiled down any future Boston urban-renewal plans to the question of "how do you raze an area without becoming some kind of monster."[30] The 1960s renewal projects in South Boston, Charlestown, and Allston-Brighton all met with profound distrust of the affected communities and with large protests.[31] If Boston wanted to renew itself, it could no longer count on bringing in the wrecking ball and building anew on the rubble. There was, however, still an area within the city limits that lacked a constituency but abounded with space: Boston Harbor.

Urban Renewal at Sea

Making a connection between Boston's fortunes and the physical shape of Boston Harbor was not a new idea. In the middle of the nineteenth century, Boston's maritime community attributed the port's commercial troubles to the harbor's shoaling, and the state introduced an expansive set of regulations that governed land making and dredging in the harbor for the next seventy years.[32] Similarly, in the first decades of the twentieth century, Boston and the state of Massachusetts tried to reverse the relative economic decline of the port by devising a series of overambitious plans for modern docking facilities that included the proposed construction of nine one-mile-long piers that would have filled Dorchester Bay all the way down to Thompson Island.[33] However, as the city's maritime commerce faltered, the harbor's perceived centrality for the city's economic well-being experienced a steady drop. During the middle third of the twentieth century, the harbor vanished not only from the sight but also from the minds of most of Boston's urbanites. It was not until the 1960s that the metropolitan gaze was directed at the harbor and its islands again. This time, however, both the perceptions and the proposed uses of the harbor differed profoundly from those of earlier times.

Instead of the trade highway and the ultimate sink of the previous periods, the city and state political representation as well as the urban planners from the BRA and the Metropolitan Area Planning Council (MAPC), an advisory body established by the state legislature in 1964 to offer a regional perspective on planning, now valued the harbor mainly for its space. The harbor, as the planners and legislators saw it, could accommodate a wide range of community needs and offered "the best chance of arresting the decline of the urban core."[34] Likewise, the harbor islands could be turned into amenities that would benefit both Boston's population and its finances. The planners found, however, that this promising space was in very bad shape at the moment. They extended the categories of disuse, decay, and blight, which they habitually applied to the urban core, to the harbor islands as well. They approached the islands as they would have any other run-down district: an area that needed to be renewed in order to become an economically productive part of the city.

For 1960s urban planners and politicians, the harbor and the islands had next to no value of their own in their current state. Representative Robert H. Quinn of Dorchester, a longtime proponent of building projects in the harbor and of making the islands "productive adjuncts of the mainland," dismissed claims that the islands may have a value other than just as real estate as shortsighted sentimentality. "We can expect some opposition from the conservationists and people who want the islands to remain forever as they are—uninhabited, unproductive, and unused. But it has always been thus," Quinn asserted. "The public good, creating a new area for expansion of the city without displacing people already there, should vastly outweigh the arguments of the bird lovers in this case."[35] Any merits the islands might have possessed, be it as breeding grounds for seabirds, sites of the last remnants of salt marshes, or homes to people, did not enter the planners' equation. For them, the harbor space gained value only by its association with Boston and by its potential to enhance the city.

Although the older dumping-ground uses of the islands were constantly vilified by the 1960s city planners and state legislators as a waste of the islands' potential, their own approach hardly differed from them in its basic quality. The nineteenth- and early-twentieth-century metropolis called on the islands to help satisfy its need for relatively remote sites to dispose of unwanted materials and people, tying the islands closer to Boston in the process and making them an integral part of the city's infrastructure. The 1960s city looked to the very same islands to solve its problems with a lack of space, a dwindling tax base, and insufficient opportunities for recreation, again regarding them as inseparable elements of the urban fabric. In 1966 the state legislature formed the Special Commission on the Boston Harbor Islands. The commission consisted of several members of the Massachusetts House of Representatives and the Senate, the presidents of the Greater Boston and Quincy Chambers of Commerce, commissioners of the state Department of Public Works and the Metropolitan District Commission (a catchall agency responsible for municipal services ranging from water supply and sewage disposal to recreation), and representatives of the governor. It was established to investigate and study the "construction of bridges, tunnels, causeways, dams and the use of land fill to connect the various islands of Boston Harbor to each other and to the mainland." The commission thus subscribed to the assumption that the definition of a useful island was an island stripped of its islandness: an island physically connected to the mainland and made into a simple extension of the city.[36]

For an example of a development embracing just this approach, the commission members only needed to look from the statehouse across the inner harbor to Logan Airport. Boston's and New England's air-travel hub had been built almost entirely on fill and was by the mid-1960s already the largest land-making project in Boston history. It had swallowed three harbor islands two decades prior, the low-lying Bird Island as well as the more substantial Governors and Apple Islands. All three islands were cut down and leveled and their earth used for filling a huge expanse of the surrounding East Boston Flats to accommodate runways, hangars, and terminals. The large granite blocks from Governors Island's historic Fort Winthrop were used as riprap at the ends of nearby runways. This formidable encroachment of the mainland on the harbor and its islands was at the same time a powerful expression of the view that state agencies would take of the harbor space in the postwar decades: a vacant area conveniently close to the city center, abounding with space and waiting to be made useful.[37]

The city yet again adopted a distinctly utilitarian view of the harbor and its islands. A 1970 report of the special commission called the harbor islands an "untapped resource" for metropolitan needs.[38] City and state agencies as well as the press repeatedly referred to the harbor and the islands as one of the most valuable "assets" of the metropolitan region, further cementing the perception of the islands as in the first place a commodity and valuable real estate. This commodity, the city fathers insisted, had been "desecrated" and misused over the years. Previous uses, they argued, had left the islands severely damaged, so that they constituted "something like a slum at sea."[39] In order to serve the city to the best of their potential, the islands and the harbor shore had to be "developed" and "improved." By improvements, the city meant altering the islands to better suit the city's perceived needs. If they were ever to become useful to the city again, they were in need of a substantial makeover.

In keeping with the perception of the islands as real estate with great potential, there was only one thing worse than using the islands and the harbor for the wrong purposes: not using them at all. Since the islands had merit only when shaped to accommodate the city's current needs, leaving them alone felt like failing to collect revenue that the city was rightfully due. In 1968 Roger Lyons, the chairman of the harbor-projects committee of the Massachusetts Bay Yacht Clubs Association, complained what a "sad waste" Peddocks Island was. Peddocks, a 113-acre island in Hingham Bay, had by the 1960s few characteristics that would qualify it as a "waste" in the eyes of most people. The hilly isle possessed a calm that the city of Boston so conspicuously lacked. It had no cars, its five connected drumlins were overgrown with vegetation, and it hosted a small colony of summer cottages. Its varied relief and pastoral-looking landscape invited activities of the less spectacular sort, like nature walks and beachcombing. To many it seemed like a welcome relic from another, quieter, time. Roger Lyons, however, did not share this view. To him, an undeveloped Peddocks was a wasted opportunity. "It has never really been a tax-producing island, nor has it ever served any useful purpose," he maintained.[40] Like Peddocks, other harbor islands that were not being intensively used to promote Boston's attractiveness and to improve its tax revenue and tourist popularity were considered "fallow" and "neglected." For far too long, a 1967 article in the *Boston Globe* claimed, had the islands lain "amoulderin' and neither public nor private interests budged to exploit so remarkable a resource." This was soon to change, however, and the islands were about to be released from their "drab condition."[41] The islands, if properly managed, could make the city more appealing to both residents and businesses and thus solve a part of its problems. By redeeming the islands, the logic went, the city could redeem itself.

To most of Boston's politicians and city planners, the harbor and its islands were vacant space, and their future could be expressed in a simple dichotomy: they could be either abused and ultimately destroyed or improved and utilized to serve metropolitan needs.[42] Either way, what happened to the harbor islands depended wholly on the preferences and needs of the city. They were regarded not so much as a natural landscape but rather as an alienated part of the urban space. The harbor did not constitute a landscape in its own right; to the city and state government, it was merely an extension of Boston.

Building in the Harbor

One incarnation of the effort to employ Boston Harbor and the harbor islands as resources for improving the city was to make them into agents of the city's aesthetic makeover. In their quest to make downtown Boston attractive to business and well-to-do residents, city planners rediscovered the waterfront setting as a potentially desirable feature of urban development. The old waterfront district, in the early 1960s still largely dominated by more or less crumbling warehouses and port facilities, was slated to become the site of modern residential and commercial development. The city's central waterfront became one of the nine principal urban-renewal projects.[43] Its proclaimed goal was to "open Boston to the sea once more." The old downtown, after the existing structures had been torn down, was to become a fashionable district of luxury apartments, office towers, and representative buildings of public education institutions like the New England Aquarium, all connected by walkways to the revitalized waterfront where pedestrian malls and boating facilities would replace the run-down piers and dilapidated warehouses. The waterfront was, according to the BRA chief, Ed Logue, supposed to become once again Boston's "Window on the World."[44] Only the world in question no longer meant foreign ships entering the city through the avenue of the harbor but rather national money settling in Boston and enjoying the harbor view.

In the effort of luring taxpayers and services, in particular the financial sector, to downtown Boston, the harbor and those islands close to the shore were to play the role of features enhancing the value and attractiveness of waterfront real estate. The MAPC praised the harbor's "scenic beauty" as a "unique urban and suburban amenity" and ascribed unparalleled scenic value to waterfront lands. "The revitalization of the Harbor," the council asserted, "offers a unique opportunity . . . to create a handsome waterfront setting for surrounding residential, commercial and industrial uses."[45] The harbor, in this line of thinking, was reduced to a reflecting pool: a pretty backdrop for Boston's potentially fashionable district.

The harbor islands' role in this way of using the harbor was twofold. On the one hand, they were supposed to serve as pleasant green adornments of the water expanse, scenic features highlighting the waterfront's attractiveness. Especially South Boston, a formerly industrial district now thought much in need of renewal, was supposed to profit from the presence of "improved" harbor islands just off its shore. "Aesthetic amenities of this sort between primary retail, office, and industrial districts," the BRA's 1965 *General Plan for the City of Boston* stated confidently, "would be quite helpful in attracting higher quality of industrial development to the Regional Core fringe."[46] On the other hand, the islands were themselves at some point expected to turn into valuable real estate and bring direct profits to the city. In 1970 the *Boston Globe* pointed out the recent development of Newport's Goat Island, a small thirty-eight-acre island in Narragansett Bay. Over a few years in the late 1960s, private investments turned Goat Island from a federally controlled site of a lighthouse and not much else into a highly urbanized seaside resort. The island lost its largely barren aspect, boasting instead an apartment complex, convention center, shopping mall, ten-story Colonial Hilton Inn, and leisure ensemble, including a golf course, tennis courts, and a 150-boat marina. As questionable as such complete takeover of an island by private development might seem, to 1960s urban planners it appeared preferable to "underusing" the available land. The *Globe* even presented the Goat Island development as an example for Boston: the Massachusetts capital should, too, take its "surplus island property . . . and then look for the most productive use, tax-wise, for the property."[47]

Private companies in the Boston area, too, seemed to think such harbor development was a viable financial scheme. In 1962 the Boston Edison Company purchased some twenty-two acres of land at Spectacle Island in the hopes of building on the island.[48] Similarly, East Coast Realty Company bought a part of Peddocks Island in the late 1950s as a real-estate investment. "We've often thought Peddocks would make an ideal residential development. There would be room for nice houses, a golf course and a marina," declared the company's president, Richard S. Robie.[49] Taking real-estate development to the harbor islands seemed like the natural next step in incorporating the harbor space into the city's renewal program.

Plans in the 1960s for how Boston could profit from its harbor space did not stop at using it as aesthetic amenity and investing in island real estate. Since the mid-1960s, predictions had been accumulating that Logan Airport would reach capacity by 1975 and that building a second airport was becoming a necessity.[50] The site most frequently mentioned in connection with this purpose was the Brewster Islands, a cluster of several small rocky islands at the mouth of Boston Harbor. Never leaving the pie-in-the-sky realm and surpassed in vagueness only by their lack of a cost estimate, these plans bore an unmistakable imprint of the era's belief in the limitless possibilities of technology. At a 1967 meeting of the special commission for the development of Boston Harbor, Representative Ralph E. Sirianni proposed linking the Brewster Islands together by four ten-thousand-foot runways that would be able to accommodate "the huge, jumbo supersonic jets of tomorrow." Passengers in this vision would have been transported between the new airport and Logan by helicopter. As proof that the proposal was very much within the trends of the era, Representative Sirianni and other members of the commission mentioned examples of other great American cities that contemplated using their adjacent water space for advancing their airport capacity. According to the dean of MIT's School of Engineering, William Seifert, both Los Angeles and Chicago were looking at the possibility of constructing airports on artificial islands off their coast or lakeshore, respectively.[51] Even New York, Boston's East Coast shipping rival of old, was reportedly considering moving the John F. Kennedy International Airport eight miles out to sea.[52]

The idea of building a satellite airport several miles off the coast was a curious mixture of a hypertrophied reliance on the power of modern technology and the not-in-my-backyard mentality. While the possibility of constructing the airport in the Brewsters area was being considered by a study group at the MIT School of

Engineering, the first impulse had come from Representative Sirianni, whose constituency was the town of Winthrop. Winthrop, along with East Boston, was the community most seriously plagued by airplane noise from the increasingly busy Logan Airport and therefore exceptionally motivated to seek the relocation of at least a part of the air traffic. Situating the airport in the seemingly empty expanse of the harbor, removed from habitation, must have seemed to Winthropites and their political representatives like an ideal solution to their problem.

More than anything else, the Brewsters airport proposal grew out of an almost complete ignorance and disregard for the harbor as an actual place. The Brewster Islands had for centuries acted as natural breakwaters at the harbor's mouth. The seas around the islands, which lay somewhat outside the enclosed area of Boston Harbor that they helped to protect, were notoriously rough and the islands themselves little more than sparsely overgrown rock outcrops exposed to harsh elements. Even the argument of the airport proponents that such a daring construction project would not be unique in the Boston region, since Logan Airport itself had to 90 percent been built on made land, did not hold on closer inspection.[53] While it was true that the vast majority of Logan's area rested on fill, the fill had been deposited atop a large expanse of very shallow tidal flats in a sheltered part of the harbor.[54] Building in the comparatively deep water around the Brewsters in the outer bay would have been vastly more difficult and also more costly. The islands' remoteness made them at the same time into popular breeding places for waterbird species such as cormorants and eider ducks; they also hosted a rare northern colony of the glossy ibis. None of these characteristics would have favored the building of an airport on the site, had they been taken into consideration. The airport plans, based on a reductionist view of the harbor as a vacant space waiting to be filled, failed to take into account the reality of the place: instead of windswept rocks surrounded by deep water and pounded by violent storms, and coveted nesting place for seabirds, the planners saw only several dots on a map signifying firm ground amid a mass of water. One of the lone voices that raised the issue of the Brewsters being in fact not very suitable as an airport site was the *Boston Globe*'s maritime reporter William Patrick Coughlin, who put the matter of the airport in the perspective of the islands' natural characteristics: "Erecting a 'Harbor Jet Airport' seaward on the remnants of the Brewster's [*sic*] one time archipelago would sum up to an investment in a dangerous location, a dreaded lee shore of wild waves in easterly storms, Boston Bay's most feared elements. Patching and leveling what's left of the surf-beaten rock-bound islands, filling in nature's created inlets . . . sounds like scupper thinking: all awash!"[55] Ultimately, nothing came of the Brewsters airport plans: by 1970 they were already, in the words of the assistant director of Massport, a government agency managing the port of Boston and the metropolitan area's airports, "as good as dead," mainly owing to the presumably astronomical cost of the project.[56] Although the Brewsters airport was put to bed, the way of thinking that conjured it in the first place was very much alive: the notion that the harbor was an even blue surface without a depth, currents, or a floor relief and without any life of its own, a vacant lot only waiting for the city of Boston to move in and build it up.

Boston's relationship to its harbor through most of the 1960s was characterized by two main traits: a special variety of technological optimism endemic to the 1960s that assured Bostonians—and, indeed, most Americans of the era—that even the most far-fetched plans were viable with the help of state-of-the-art technology, and the effort to use the harbor as intensively as possible to solve Boston's urban problems.[57] They both manifested themselves most conspicuously in the most ambitious of the unrealized harbor projects of the 1960s: the World's Fair of 1976 and the new harbor community.

Expo 76 and the New Community

Ideas to use parts of the harbor for building a residential community began surfacing as soon as the U.S. Army released several of the harbor islands from its ownership in the late 1940s. By the early 1960s, with the Boston urban renewal well under way and with the central city looking for ways to boost its appeal to the middle class, such voices were becoming louder.[58] In particular, Dorchester representative and house majority leader Robert H. Quinn was advocating the filling of a substantial portion of water between the islands closest to the city and using the made land for commercial and residential purposes.[59] For two years, Quinn's proposals did not find much resonance; by his own account, he "felt like he was talking into a rain barrel."[60] It was only when the proposal of a harbor community was merged with the plan to hold a world's fair in Boston in 1975 that the idea started to catch on.

In 1962 Gilbert H. Hood, president of Hood Dairies and also of the Greater Boston Chamber of Commerce, led a delegation of the metropolitan region's business leaders on a visit to the world's fair in Seattle. The businessmen were impressed: after their return, they recommended to Mayor Collins that Boston should host a similar event. To them, such a fair represented mainly a business opportunity: they cheerfully asserted that the fair could bring "$500 million in retail sales" to the region's economy.[61] What started as a relatively vague idea about boosting sales soon grew into a concerted effort to bring the next international exposition to Boston and to combine it with the nation's bicentennial celebration. In 1963 the Chamber of Commerce formed the nonprofit corporation 1975 World Freedom Fair, Inc., or simply Freedom 75, for the purpose of securing the fair for Boston.[62] The corporation's board of directors was composed of sixty New England community leaders, including six state governors and several university presidents. State and city agencies, including most prominently the BRA, as well as urban planning experts from Boston and Cambridge universities cooperated with Freedom 75 on the planning process. After meeting with state and federal legislators, local planners, members of the corporation that planned the 1964 World's Fair in New York, and officials from the Bureau of International Exhibitions, Freedom 75 publicly announced its intent to hold the bicentennial celebration in Boston.

Perhaps prophetically, they did so on November 9, 1965—the night of the Great Northeast Blackout that plunged eight northeastern states into several hours of darkness. They did not, however, see the writing on the wall just yet. As the BRA annual report noted, "The lights stayed on long enough for the proposal to win the community's continued enthusiastic support."[63]

Apart from the nation's two hundredth birthday, the fair was also meant to celebrate the completion of Boston's urban renewal.[64] Ideally, it would have not only shown off the achievements of Boston's revitalization but also added a whole new part of town that the visitors could admire. Temporary pavilions of the world's nations were to be complemented by a permanent community that would enhance the city's housing, commercial, and recreational potential. The international exposition would thus showcase Boston's new liveliness and attractiveness to the whole world. As Freedom 75's planning coordinator, Martin Adler, put it: "We will have a new city within the city to show the world at our country's birthday party. It would be a birthday cake for the world, and the celebration would be the icing on the cake."[65]

From the beginning, almost all of the sites under consideration for the exhibition were part of the harbor. The former Boston and Maine railroad yards in Charlestown, East Boston piers, and South Bay were all contemplated as fair sites in the first planning stages, all of them on the city's waterfront.[66] The choice was a logical one: the former industrial and commercial districts were among the few areas in the inner city that had anywhere near enough space to accommodate the fair, and their use for exposition purposes would have turned the city's bothersome "blighted areas" into modern development, paid for largely by federal and international money. It was, however, the harbor's "underused" islands that attracted the fair planners the most. They combined the practical advantages of available land relatively close to the city center with being "unusual, scenic, and topographically interesting."[67] The planners thus picked up the widespread narrative of "taking advantage of a long neglected asset on Boston's doorstep" and reiterated the widely held opinion that the city as a whole would profit from converting the islands into "productive urban areas."[68]

The islands offered a unique and, from a cultural history perspective, fitting setting for the planned world's fair and the new community. The traditional perception of islands as places apart from the mundane reality of the mainland, as small, self-contained universes unburdened by the problems of continental life, fed right into the utopian quality of the fair project.[69] All previous world's fairs on the American continent were, in the words of historian John Opie, "fantasy worlds incarnate" and presented Americans with an "explanatory blueprint of the future."[70] The proposed island site of the fair reinforced these qualities by physically separating the exposition-to-be from the pedestrian world of the present-day city. Similarly, the promise of a new urban way of life in the harbor community had a decidedly insular character of the utopian variety. The planners of the fair acknowledged this explicitly: David Harrison, who was assigned by the BRA to work with Freedom 75 on the fair project, wrote in 1967:

> The city of tomorrow on its island site will be physically as well as conceptually a world unto itself, impressing upon its inhabitants and visitors its stature as a place not to be compared with any other, a place to come to for a purpose, and never to be taken for granted. Between it and the mainland there will be a border; to cross it will be an experience in itself, a break with something that has gone before and a looking forward to what will come next.[71]

To escape the problems of extant cities, the community of the future needed to be separated from them spatially as well. In the Western cultural tradition, this was most easily accomplished by locating this un-city-like city on an island. Through their site choice, Boston's urban planners stressed how different this new community would be from the old, blighted urban core.

By siting its fair on the harbor islands, Boston was about to copy Montreal, the site of the expo of 1967. Throughout the 1960s, the planning and construction of the Montreal World's Fair functioned as a catalyst of the transformation and renewal of that city's infrastructure and image, marking Montreal's ascension as a "modern, world-class metropolis."[72] Even the *New York Times* acquiesced that the exposition transformed Montreal into a "swinging cosmopolis," although the paper found it hard to extend the potential effects of a similar fair on Boston or its rival Philadelphia: "The mind boggles at the thought of staid Boston or stodgy Philadelphia turning into exciting, captivating world tourist centers."[73] Not surprisingly, Boston's leadership took a more positive outlook on their city's future gains from the world's fair. When a fifty-four-member Boston delegation, led by Mayor Collins, visited the Montreal Expo in June 1967, the question on everyone's lips was: "If Montreal can do it, why can't Boston?"[74]

As Boston's planners were meaning to do, the Montreal Expo used the makeover of the "decayed heart of the city" on the St. Lawrence River waterfront to rejuvenate and modernize the whole town. The Montreal predecessor to Boston's community of the future was Habitat 67: architect Moshe Safdie's radical experiment in modern urban housing situated on the St. Lawrence River bank. Its purposes were manifold: redeeming the "blighted" waterfront district, serving as a prototype of what life in cities would be like in an ideal future, and "remarrying the city to its waterfront" through opening a view on the river and the Montreal harbor. Most of the Montreal Expo activity concentrated on the St. Lawrence River islands: the city of Montreal provided St. Helen's Island for expo purposes and also created a new island, Île Notre-Dame, in the stream to host the exposition.[75] Both Montreal and Boston combined in their treatment of their islands the somewhat contradictory historical notions of the islands' nature: they regarded them as places apart, and therefore eligible for special uses; at the same time, the islands were treated as extensions of the city and as its representatives, as places where the city could heal and reinvent itself. In doing so, both cities in fact replicated the treatment of urban islands that had been prevalent since the nineteenth century: their perceived otherness was combined with a close functional attachment.

The first proposals for the Boston World's Fair, or Expo, extended the future city far into the harbor. Some of the earliest plans envisioned a line of fill and construction that extended as far as Lovells and Georges Islands or even the Brewsters. In 1966 Freedom 75 put forward a plan that included four principal island sites: Thompson, Spectacle, Long, and Moon Islands. Spectacle's size and central position seemed to make it into a natural focal point for the exhibits, and later, it was proposed, it would make a fine site for regional community facilities. Even though their final choice was an island closer to the shore, the planners nevertheless noted that the more remote harbor islands would be "both adequate and unusually exciting sites for the Bicentennial."[76]

Eventually, the proposal shrank to just Thompson Island and its mainland pendant, Columbia Point. Thompson Island, a 170-acre island in Dorchester Bay, had been home to various institutions caring for orphaned and disadvantaged boys since 1833. Only a narrow water stretch separated it from the nearest shore of Squantum Neck, and the island could be reached on foot over a sandbar at extreme low tide. Both the island's relatively large size and its proximity to the mainland invited plans for its development. Columbia Point, a 350-acre peninsula opposite Thompson Island, was by the mid-1960s an embodiment of social isolation and neglect. It was the site of the Calf Pasture sewage-pumping station, several closed landfills, and a housing project built in the early 1950s that was already gaining notoriety as a symbol of all that was wrong with postwar public housing projects. To the Freedom 75 people looking for a scenic site close to the city center, whose redevelopment would showcase the achievements of Boston's urban renewal, the combination of Thompson Island and Columbia Point represented an optimum choice. The plan was to develop the site for two purposes: First, the bicentennial would be held there, an exhibition that would present the booming face of new Boston to the world, among other benefits such as increasing tourism and creating jobs and revenue. After the exhibition's close, the land would be reused for a residential community that would at last fully integrate Thompson Island and Columbia Point into Boston's built environment.[77]

Realizing this vision would have meant a complete subordination of the site's environment to the exhibition and new community purposes. In the best Boston tradition, much of the new development was supposed to be built on fill: six hundred acres of new land were to be made between Thompson Island and Columbia Point. Tunnels and viaducts would have connected the island and the mainland, and a dam was proposed between Thompson Island and Columbia Point to create a constant-level boat basin. Even ostensibly environmentally friendly measures like alleviating water pollution were motivated by making the site appealing to future residents. The planners interpreted Thompson Island's natural features—its hilly terrain, sea breezes, position in the harbor—exclusively as "scenic amenities," making it into attractive real estate. Even its thirty-acre tidal lagoon, from today's point of view a visually striking and ecologically valuable wetland, was regarded by them as mainly the preliminary stage of a marina.[78]

Freedom 75 stressed frequently how unique and unusual the location was; however, the early expo design certainly was not. Its 1966 visualization could have been taken straight out of the design book of the 1964 New York, the 1962 Seattle, or indeed the 1967 Montreal World's Fair: a strip of futuristic pavilions stretching into the distance, complete with the inevitable "space needle." The suggested program also seemed to emulate the traditional fare of such events: the general themes included "scientific exploration and discovery," "physical resources," "human culture," and also a "Merry Universe" exhibit, ideally featuring amusements such as "large replicas of famous sights, like those now on display in Copenhagen's internationally famous 'Tivoli Gardens.'" Despite its professed goal of harnessing the resources attracted by the international exhibition for the good of Boston and its residents, the design seemed completely divorced from the reality of the city.[79]

FIGURE 17. An early visualization of the Boston World's Fair envisioned intensive development and filling of a portion of Dorchester Bay. The design itself was typical of the early 1960s, complete with a "space needle."—*Image from the 1966 report* Freedom 75: The United States Bicentennial Celebration—a World Freedom Fair in Boston.

Not everyone liked the direction in which the expo planning was going. The young architect Jan Wampler, a Harvard School of Design graduate who had spent the previous several years in South America and Puerto Rico, came back to Boston in early 1968 and thought the expo plan was "terrible." To him, it failed to realize the potential of an international exhibition to seek solutions to common problems of mankind on a global scale. He had nothing but contempt for the 1964 New York fair: "A great big one billion dollar party at the expense of millions of people whose basic needs are ignored. . . . What a mess!" Wampler approached Ed Logue with the offer to develop an alternative idea for the exposition, which by that time had been pushed back to 1976. Logue liked the plan, and twenty-eight-year-old Wampler was hired as chief designer of Boston's expo. He assembled a team of young architects, most of them recent graduates of Boston and Cambridge universities like himself. The "Planning Design Group" set up shop in a run-down building on the waterfront and spent the next several months practically living together, playing the Beatles' records, and working on the design of a new kind of expo: one that would help solve urban problems not only in Boston but everywhere.[80]

The expo plan that Jan Wampler's team put forward was a bold departure from the way such expositions had been understood in previous years. Instead of offering individual countries a vehicle to show off the achievements of their national science and industry, the Planning Design Group envisioned a problem-oriented expo with participants working toward common solutions to problems of the contemporary world. In Wampler's eyes, one of the most pressing universal

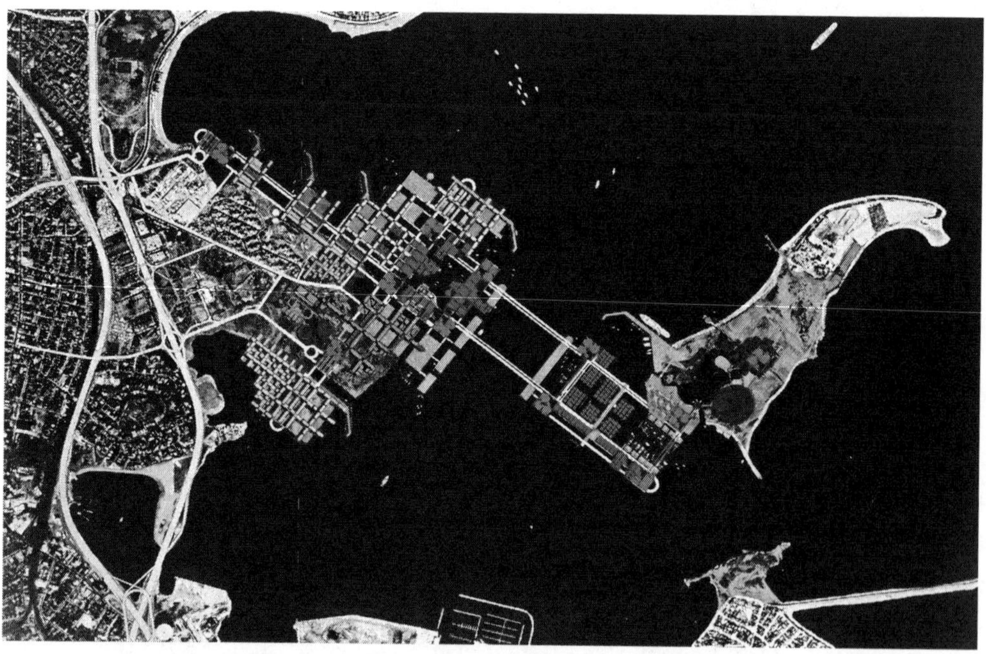

FIGURE 18. Jan Wampler's 1968 plan of the expo. A large part of the exhibition would have been built on fill and floating platforms between Columbia Point and Thompson Island. —*Courtesy of Jan Wampler and the Planning Design Group.*

issues was urbanization and the social, environmental, and economic problems that came with it. This was, of course, not just his view: in the United States of the 1960s, urban issues were in the center of academic and political debate. New urban-design graduate programs promoted a holistic view of city planning, federal funding was available for urban-renewal programs all over the nation, and theorists like Donald Appleyard, who taught city planning at MIT in the early years of the decade, emphasized the essential role urban environment played in the lives of city dwellers.[81] Urban problems were both local and global: they were problems Boston shared with hundreds and thousands of cities all around the world. If the Boston Expo could help tackle urban problems of pollution, traffic congestion, urban sprawl, social inequality, and racial segregation, its designers argued, these results could not only improve life in the Hub itself but also serve as a blueprint for other cities contending with similar issues. Accordingly, the expo was supposed to fulfill two fundamental functions: as a temporary urban laboratory, it would help solve common problems of an urbanizing world, and as a permanent new community for Boston, it would put these solutions into practice.[82]

The site choice and design corresponded with these goals. Located just minutes away from the urban core, the developed Columbia Point was to become a direct extension of the central city and an alternative to suburban sprawl. The new community for fifty thousand people would offer mixed-income housing and rely on mass transit and pedestrian walkways, in direct contrast to the car-dependent and socially homogenous suburbs. The new town was also supposed to make the most of its harbor setting. In accordance with the views promoted by city and state agencies such as the BRA and the MAPC, the young designers saw the waterfront and the harbor as assets that the city should make productive use of. Boston was just one of many major American cities located on waterways, Wampler and his team reasoned, and if it utilized the space of the harbor for construction, transport, and leisure, the city could become a pioneer of a new urban way of life. The design itself made clear that its authors regarded Boston Harbor as an essentially urban space. The project's striking visualizations showed high-rise buildings, pavilions, bridges, escalators,

FIGURE 19. The Water Plaza was supposed to become the core of the expo and, subsequently, of the new community. The plan understood Boston Harbor as essentially urban space.—*Courtesy of Jan Wampler and the Planning Design Group.*

and sidewalks suspended over the harbor waters. Built on existing waterfront and island, as well as 465 acres of fill and floating platforms, the expo and the new community would have meant the ultimate expansion of the city into the harbor.[83]

The urbanization of Thompson Island took a form that emphasized both its integration into the urban network and its quality as open space. While working on the project, Jan Wampler spent a night out on the island in order to get a sense of the place. When he later put into words what struck him about the island, he stressed the ambiguity of its position vis-à-vis the city as well as its potential for recreation. "When you are on the island," he wrote, "you sense its closeness to downtown Boston and at the same time its remoteness from the general chaos of urban congestion—a great paradox, like Central Park in a way." Accordingly, it was crucial both to make the island accessible for the community and to preserve its natural beauty. The Thompson Island design reflected these in fact contradictory tendencies. Although most of the island's 130 acres were reserved for a natural park with a youth camp and a demonstration farm "to acquaint urban young people with the world of nature and growing things," the island's southern end would also accommodate a "Recreation Theme Center," a marine research institute, a five-hundred-boat marina with resort hotel, and even a Buckminster Fuller–esque "Transparent Dome," a 384,000-square-foot environmentally controlled sphere meant to make picnicking possible in every weather and season. The notion of intensive development as the preferred way of productively using the harbor and its islands was here somewhat awkwardly married to a vision of the harbor as Boston's greatest open space and recreational asset.[84]

The expo plan was but a moderate expression of a general tendency of most contemporary politicians and city planners to see the harbor as a space Boston could, and eventually would, expand into. Jan Wampler saw his new community as the nucleus of a

FIGURE 20. Thompson Island was slated for recreational use, part of which would be a demonstration farm. In the background is the environmental-control dome for year-round picnicking.—*Courtesy of Jan Wampler and the Planning Design Group.*

future satellite core of Boston's metropolitan region. The harbor community would be the focal point of Boston's expansion to the east—that is, into its harbor. The harbor islands, as he saw it, would be incorporated as open space of a new harbor neighborhood serving a population of 150,000 people by the year 2000. Some of the other harbor studies undertaken in the 1960s were even more radical. A group of architecture and planning students at Harvard predicted that if the city's development continued as previously, by 2040 there would be no more Boston Harbor. The entirety of it would have been filled with tunnels, causeways, dams, and fill, paved over to make way for the needs of the expanding city. Even though unrealistic, such plans spoke volumes about the prevalent 1960s imagination of Boston Harbor as free space where a new and better Boston could be built.[85]

Even at the time, this approach to the harbor and its islands as dormant facilities at the city's command was not universally accepted. James B. Ayres, a *Boston Globe* reporter who made a name for himself with pioneering environmental coverage in the early 1960s, criticized the "got-a-problem" mentality with which city officials approached the harbor in a piece titled "Time for a Change: Boston's Harbor—a Problems Dump." In it, he subjected the city's way of thinking about the harbor to scathing criticism. "Boston officials," he wrote, "look to Boston Harbor whenever they have problems. Got a problem? Sweep it under the water of Boston Harbor. . . . Got a trash disposal problem? Burn it on Calf Pasture, Spectacle Island, Gallups Island or in barges off Calf Island. . . . Got a low-income housing problem? Build it on Columbia Point and isolate the people from the rest of Boston." The expo proposal, he suggested, was no different in its underlying understanding of Boston Harbor as a space that existed only to solve Boston's problems. "Got a space problem for living, working and playing in Boston? Build causeways, dams and tunnels to connect the islands of Boston's inner harbor. Got a problem with urban renewal displacing people from their homes? Build bridges or dump fill between the islands of Boston's inner harbor."[86] Instead of externalizing the housing and recreational space problem in the same way that nineteenth-century Boston sent away its trash, he argued, the planners should look to fix the city they already had.

Others were concerned about the potential effects of the world's fair project on water pollution in the harbor. The project was accompanied by a number of studies that assessed its potential impact on harbor ecology and on tidal flow in Dorchester Bay, and the BRA and Freedom 75 planners claimed that it would be part of the expo plan to abate pollution.[87] However, residents of shore communities voiced fears that exactly the opposite might happen: the extensive amount of fill would, in their view, have disrupted water flow in the harbor and made the pollution problem even worse.[88] The expo proposal also threatened to clash with other uses that, in themselves, were just as one-dimensional in their perception of the harbor. The Massachusetts Bay Yacht Clubs Association organized a "conservation cruise" around the harbor in 1968 and another in 1969.[89] During the first one, more than one hundred boats gathered to do a loop around the harbor and draw attention to the yachtsmen's agenda. Despite the cruise's name, the yacht clubs' demands had less to do with conservation and more with shaping the harbor up for yachting. Among their requests was a new public pier downtown, the construction of new marinas along the shore, more dredging, elimination of floating debris, and more and better public access to the harbor islands.[90] To city planners, the harbor may have been a combination of a vacant lot and a scenic element to adorn their constructions; to the yacht-club members, it was a recreational facility serving first and foremost the interests of boating. Despite finding themselves on the opposing sides of the world's fair project, they shared a basic utilitarian understanding of the harbor as a useful adjunct to the city and its needs.

Local opposition ultimately proved a more deadly foe to the expo plans than the competition from Philadelphia. Senator John Joseph "Joe" Moakley of South Boston dismissed the expo plans as "pie in the sky" and also questioned, besides their impact on the tidal currents and harbor pollution, the plans' financial soundness and overall practicality.[91] The issue also saw the involvement of the conservative mayoral candidate of 1967, Louise Day Hicks, who came down heavily on the side of the expo opponents. As president of the South Boston Residents Group, a local civic organization, she led vocal protests against the fair and, in the words of the *Globe*'s longtime harbor correspondent Ian Menzies, "attacked Expo 76 with the same vigor as Washington took Dorchester Heights and held her position equally tenaciously."[92] Claiming concerns about harbor pollution and the heightened tax burden that the expo would mean for Bostonians, Hicks, a known opponent of desegregation and court-ordered busing, also fought against the potential establishment of a mixed-income and minority neighborhood in the vicinity of the predominantly white South Boston.[93] Although a poll done by the *Boston Traveler* showed most respondents in favor of the fair, Hicks and Moakley still managed to get South Boston opponents of the expo to show up en masse at a Boston City Council hearing, carrying pickets saying "Don't Be Silly—Send It to Philly!"[94]

Finally, it was the lack of local government support that broke Expo 76's back. The Boston City Council in September 1969 unanimously voted against the expo plan, judging the proposed site "unsuitable" and the whole project "not in the best interest of the City of Boston."[95] This, in turn, prompted Jan Wampler to quip: "If we had a City Council around 200 years ago, we would not have had an American Revolution."[96] Since the city had been expected to pick up a substantial part of the tab, $75 million out of the $500 million expected to be paid from the public purse, the state followed suit, Governor Francis Sargent's spokesman declaring: "That about does it. I don't see how the state can support [the fair] in the face of Boston's opposition. It looks very grim at this point."[97] A Boston delegation nevertheless presented its official bid in Washington only weeks later. Along came, however, not only "several 'Bedford Minutemen,' wearing three-corner hats, carrying muskets and beating drums," but also Senator Moakley and Louise Day Hicks, who carried their criticism of the planned event even into the congressional hearing rooms.[98] Unsurprisingly, given the lack of secured financing and a feud within the Boston representation, the bicentennial celebration went to Philadelphia instead, effectively canceling Boston's plans for a world's fair. All that remained was a model of the expo and of the new community, hidden away in Room 917 at Boston city hall.[99]

Although not realized in the end, the expo project opened the door for seeing the harbor and the waterfront as a potential development site where the city could reinvent itself and try to solve its problems. There was one part of the Planning Design Group's proposal that was eventually realized: the University of Massachusetts in Boston relocated its campus to Columbia Point between 1968 and 1974. The UMass campus was one of the first uses of the harbor shore in a long time that did not regard it as a fringe location best suited for unsavory businesses and social isolation of the

FIGURE 21. The University of Massachusetts campus was part of the expo and new community plan intended to improve Boston's waterfront—and the only part of the design that was actually built. The Calf Pasture pumping station and the Columbia Point housing project are visible behind the UMass buildings. —*Courtesy of University Archives and Special Collections Department, Joseph P. Healey Library, University of Massachusetts, Boston: UMass Boston: Columbia Point Collection.*

poor. Rather, it was a brainchild of a way of looking at the harbor and its shores that the expo started: as a potentially attractive space for reinventing and enhancing the city.

Developing the Harbor for Recreation

The argument most often used against the expo was that the exhibition would interfere with the harbor's potential for recreation. South Boston residents wrote angry letters to the *Globe* editor, decrying the allegedly imminent obliteration of the neighborhood's popular Carson Beach and claiming that nothing less than "the rape of our beaches" would result if the expo project were allowed to go forward.[100] Others formed a "group of concerned citizens" from the communities around Dorchester Bay, united under the motto "Save Our Shores," or SOS for short.[101] These groups juxtaposed the planned building projects in the harbor with a notion of the harbor as a supposedly exclusively natural space that would be "desecrated" by uses other than outdoor recreation. "The harbor is natural," Virginia Sheehy, the former secretary of the South Boston Residents Group, declared. "Let it stay that way."[102] The harbor recreation boosters attempted to establish a hierarchy of uses in which the harbor's function as a recreational area was ascribed the highest value. However, even though the words *conservation* and *nature* were dropped casually in the conversation, the reality of most proposals was far removed from conservation of any kind. Rather than by concerns about the health of Boston Harbor ecology, many of the 1960s initiatives that advocated conservation of the harbor's natural resources were fueled by the vision of developing the harbor as an urban playground.

Recreation in Boston Harbor could look back at a considerably shorter history than land making, but by the 1960s it nevertheless had a tradition that spanned more than one hundred years. In the heyday of shipping commerce, recreation mostly took the relatively unobtrusive form of boating trips that had low demands on shore space and facilities. In the 1840s, even Spectacle Island was a popular destination for such outings. It had two summer inns built on it, and parties from Boston would come over for picnics and clambakes.[103] In the course of the nineteenth century, yacht clubs sprang up around Boston Harbor: in 1885 there were fourteen of them, and at the height of the season seven or eight hundred boats moored at the City Point Yacht Club alone.[104] In the 1870s and 1880s, Boston Harbor was understood by Bostonians not only as a place of labor but increasingly also as a place of leisure and beauty. Steamers plied the harbor waters between the downtown Foster's Wharf, City Point, and favorite summer resorts down the shore such as Lovell's Grove in Weymouth or Melville Garden at Crow Point in Hingham. The latter, established in 1871 by Dorchester industrialist Samuel Downer, was a particularly popular resort, offering the full catalog of pastimes expected in an amusement park of the era: bowling and shooting alleys, swings, rowboats, a restaurant and music hall, daily clambakes, a monkey cage, and a Punch and Judy show.[105] A boat ferried guests to the nearby Ragged Island, where they could enjoy the view from a lookout pavilion built atop the island's rocky elevation. Although the harbor was not the park's main feature, it added significantly to its popularity: most patrons reached the resort via a pleasant steamboat ride down the harbor from Boston, and harbor views were listed prominently among the park's attractions. "There is no more beautiful sight on a sultry day," mused the author of an 1879 Boston Harbor guidebook, "than to sit under the veranda of one of the beach-houses, and watch the yachts sailing with their snowy sails over the blue waters of the bay, bending gracefully to the freshing breeze."[106] At other spots around the harbor, numerous hotels promised accommodations and fish dinners to yachtsmen, transforming the harbor into a water park where a most pleasant time was to be had, if one had but money to buy a boat.

In the late nineteenth century, urban masses who indeed could not afford the financially demanding pastime of yachting came to enjoy the shore as well. Convinced about the positive effects of fresh sea air on both the physical and the moral health of the urban population, the city of Boston started acquiring and developing the shoreline for the benefit of its less well-to-do residents. In 1881 South Boston's City Point Improvement Association began to contemplate building a marine park to benefit the people "who could not ride out into the country in carriages" and who "had no other place near at hand where they [could] get a breath of fresh air."[107] Although to a large extent driven by a profound middle-class uneasiness about the potential threat that the immigrant urban poor posed to social stability, the effort to replace "vicious amusements" like dance halls and saloons with "wholesome recreation" opened the harbor to all classes of Boston's residents as a place of recreation.[108] The City Point marine park, designed by the star landscape architect Frederick Law Olmsted, opened in the 1890s and proved immensely popular, with an estimated visitation of 40,000 people on summer Sundays around the turn of the century.[109] Regional seaside resorts and beaches were equally well visited: the ocean beaches at Nahant, Revere, and Nantasket at the Boston Harbor entrance were among the most coveted recreational sites in the metropolitan area and drew large crowds during the summer season.[110]

By the turn of the century, the first golden era of Boston Harbor recreation was slowly becoming history. Melville Garden in Hingham closed in 1896, its land sold for residential property; the ferry services in the harbor were gradually discontinued during the first half of the twentieth century, as the city's and state's attention shifted toward constructing automobile infrastructure on the mainland. The ocean beaches continued to be extremely popular among Boston residents, with a reported 150,000 people visiting Revere Beach north of Boston on a sunny day in late June 1930.[111] The city, however, no longer saw harbor recreation and investments in its infrastructure as its priority. Sea air, strolls along the shore, and ocean vistas were more relics of a bygone time than elements seen as necessary for the health of the urban population; they had long since been replaced by more active pursuits. Developing sites that offered "health-giving sea breezes" and "charming views of harbor life" had been curbed in favor of swimming pools, ski slopes in the Blue Hills, and skating rinks.[112]

The postwar decades, however, saw a renewed surge of interest in seaside recreation. Urban-renewal planners all over the country worked not only on transforming run-down urban cores into fashionable and tax-producing districts but also on transforming what they saw as "blighted" urban beaches into modern waterfront playgrounds, contributing to the overall structural and aesthetic makeover of central cities.[113] More important, recreation by the sea was rediscovered as a potentially tremendous source of revenue. All contemporary studies on open space and recreation agreed: never before in the history of the nation had leisure been so important to the economy. The postwar United States had an exploding population with more money to spend, more leisure time on its hands, and more mobility available to it than any other generation before them. These affluent, mobile, and leisure-seeking people were apparently desperate for recreational opportunities—which, as all the studies agreed, the Commonwealth of Massachusetts was too slow in providing for them.[114] A report commissioned by the Massachusetts Department of Natural Resources (DNR) in 1955 asserted that Massachusetts could profit immensely from its natural resources and was wasting the opportunity. Above all, the state's two thousand miles of seacoast were cited as an untapped resource that could boost the Massachusetts tourist industry. As the report claimed, the coast was a promising recreational resource sought by tourists from both in and out of state: "There is a natural affinity between recreation and beaches. People want to get near and into the water." The coast's recreational potential, according to the DNR report, surpassed by far the appeal of the more traditional attractions of the Boston region, which were mostly connected to the city's historicity. Modern recreation was supposed to have less of an educational and historical focus and concentrate on providing visitors with an opportunity to enjoy activities, most prominently sports, in a natural setting. As the authors of the report somewhat bluntly put it: "They [sightseers] are not likely to make repeat visits merely to see Paul Revere's house or the Minute Man a second time. They will be more apt to come again if they had a good time at the beach, or a lake shore, or in some other pleasing recreational environment."[115] This recommended "pleasing recreational environment," however, was not simply there for the taking. In the view of state and metropolitan agencies, it needed to be developed and improved first in order to accommodate the supposed tastes of the visitors, if any good time was to be had at all.

In the Boston metropolitan area, this perceived need for development of ideal recreational settings turned toward both constructing resorts that conformed to modern trends in leisure and architecture as well as the preservation of what the planners called "regional heritage." This was in fact the stereotypical image of a New England landscape of elm-shaded village greens, stone fences, steepled churches, and white clapboard houses. One of the cornerstones of this New England open-air museum was the fishing village, "very much in the Nantucket or Marblehead tradition, with . . . closely built houses on winding, cobbled lanes leading down to a bustling but tidy harbor."[116] This picture, the planners admitted, was "difficult to find still existing in the study area," mainly because, as John Gillis has pointed out, it never actually existed. The peace, tidiness, and timelessness of ancient fishing villages were an invention of nineteenth-century urban visitors that belied the transience, instability, and inherent modernity of actual coastal settlements.[117] If visitors were to find romantic fishing villages huddled on the Massachusetts coast, almost as much development would have been needed as with the construction of state-of-the-art recreational facilities.

Similarly to maintaining colonial kitsch along the coast as a tourist magnet, state and metropolitan area agencies saw Boston Harbor and its islands as potential urban recreational facilities that needed only to be developed as such. At a time of a perceived growing demand for leisure, the harbor and islands, simultaneously to being contemplated as resources for residential development, became a "major asset for recreational use."[118] Whether as real estate for a new community or as potential recreational resort, the islands were seen as an urban amenity and an indivisible part of the city's fabric. "Open space in an urbanizing world is a direct service function," the MAPC's *Open Space and Recreation Plan for*

Metropolitan Boston stated in 1969, quoting an earlier position paper of the American Institute of Planners. "We do not see open space standing by itself, an island of green in an ocean of urban development. We see it as a part of a 'systems' approach to the social and economic requirements of the American people."[119] For 1960s city planners, the harbor and islands were not set apart from the rest of the city's space; they were yet another blighted district that needed renewing and could be put to use and work for the city's and its population's social and financial benefit. The harbor space was fulfilling a certain function in the city's infrastructure, contributing to the overall functioning of the urban system.

Of the proposed uses of the harbor islands, recreation was presented by its proponents as the most "natural" use and as synonymous with conservation and "saving" the harbor. Most plans for recreational use of the islands, however, involved almost as much dredging, filling, and building as the hypothetical, and much vilified, construction of the expo. One of the many recreation plans for the metropolitan region, brought forward in 1967 by the MAPC, called for a scenic highway from Winthrop to Deer Island where a marina would have been built. Spectacle Island was to be reconstructed and converted into a picnic area; the plans also included the construction of a small artificial island off Carson Beach to protect the beach and double the space for swimming.[120] In its impact on the harbor environment, the most prominent recreational proposals of the 1960s, if realized, would have equaled the envisioned residential development of the harbor. Like the planned new community, the recreational development was best described as urban renewal at sea.

As was the case with its expo, Boston looked to other cities for inspiration on how to turn its islands into a recreational paradise and "productive adjuncts of the mainland."[121] The Canadian province of Ontario, which devoted several newly formed islands in the dammed St. Lawrence River to leisure purposes, according to the *Boston Globe*, offered guidelines for the recreational use of the Boston Harbor Islands.[122] Even more relevant seemed the example set by Toronto: the Canadian port developed a group of small islands off its lakeshore into a water amusement park, only a ten-minute boat ride away from downtown. Linked by "Venice-like" bridges, the islands offered amenities ranging from picnic tables, grills, and restaurants over pony rides and merry-go-rounds to pedal boats and regatta courses. A reported thirty-five thousand people a day boarded the ferries to the islands and spent a pleasant day frolicking on beaches and strolling along the boardwalks of the urban water park.[123] Here, the quality of islandness was turned into a mere attraction: the novelty of being surrounded by water in an otherwise completely urban setting.

Boston's most ambitious proposals for the recreational use of its harbor islands would have surpassed even Toronto's intensive development of its insular park. In 1969 the Harbor Islands Study Group at MIT published a report commissioned by the legislature's Boston Harbor Islands Commission. In it MIT students and faculty presented three options of harbor development, all of them resting on the basic premise that "Harbor development may offer the best chance of arresting the decline of the urban core." Although listing "conservation and recreation" as one possible alternative of developing the harbor islands, the study group emphatically recommended the other two options the report presented: a harbor community, a plan almost identical with the BRA and Freedom 75 proposal, and a plan for an intensive recreational development of the harbor islands. In this alternative, the islands were to become year-round recreational and amusement appendages of the city. They would have formed "natural extensions" of the mainland, both in form and in function: in visualizations, the islands, connected to the mainland by bridges, fill, or at least frequent ferries, were reaching like fingers from the shore deep into the harbor. The shores opposite the islands would have become landward bases of recreation, with transportation hubs, high-density residence, and public marinas providing cheap access to the islands for all social classes. Long, Thompson, Peddocks, Moon, Grape, and Slate Islands would have all featured major beaches, with additional facilities for camping, cycling, boating, hiking, and other types of outdoor activities, while also including amusement parks, theaters, hotels, and apartment complexes. Three of the islands—Lovells, Gallops, and Georges—the plan would have linked to one another via pedestrian bridges, built them up with hotels as well as beaches and marinas, and made them the "rendezvous for East Coast yachtsmen and tourists." Beside ferries and yachts, transportation between the islands and the mainland was supposed to be secured by a system of small buses and aerial tramways.[124] Although planning to preserve some of the islands in their "natural" state, this proposal quite clearly understood the harbor islands as a simple extension of the city, with the potential to ease some of the problems of the metropolitan area, in this case the demand for recreation and open space. When the Metropolitan District Commission acquired Peddocks Island in 1970, it adopted the same basic understanding of the island as empty space existing to serve the community's needs. When purchased, the island was characterized as "139 acres surrounded by salt water, only eight miles from City Hall." The island was defined by its size, allowing for a variety of intensive recreational development, and by its proximity to the central city that enabled Boston urbanites to use it as a resort. Utilizing the islands remained the city's priority. "Peddocks, it is planned, will be a recreation area for Bostonians. And some day the harbor will be cleaner, and some day something will be done to utilize, for fun or more beauty, the other islands in what should be one of the world's most attractive waterways."[125] What these plans ultimately amounted to bore a baffling similarity to the expo's new community. They foresaw tying the islands as close to the city as possible both physically and functionally and subjecting their environment completely to the needs of the city. The harbor recreation plans proposed in the 1960s envisioned nothing less than turning the harbor into the city—only a better one.

The Future of Boston Harbor

Boston Harbor as the site of a new community, recreational development, or a seaport and airport: all these potential and actual uses were discussed at a high-profile conference titled "On the Future of the Islands, Waters and Shoreline of Boston Harbor" that took place on June 24–25, 1969, at the Massachusetts State House. Organized by the Boston Harbor Islands Special Commission, the conference brought together federal, state, and local government officials, academics, and representatives of private industries to contemplate the future of the harbor. In nine workshops over two days, delegates probed the harbor's and the harbor islands' potential for the city. All the familiar uses of the harbor space were revisited: in separate workshops, the harbor's role as a seaport and a waterway, its use for waste disposal and the connected problem of pollution, as well as its potential for recreation and as a possible site of a harbor community were all discussed.[126]

Although seemingly considering myriad approaches to the harbor space, the conference had only one underlying conception of Boston Harbor and the islands: as an urban facility. Conference participants regarded the harbor as Boston's "salvation," a panacea for all kinds of urban ills. To them, the harbor was "open space, usable space," a multipurpose facility able to accommodate, when managed properly, a multitude of urban uses.[127] At the same time, it hardly seemed to have a life of its own that would have been worth protecting or even considering. Although a panel on harbor ecology was part of the conference, it remained focused on "recoverable resources" that could be developed and utilized; to the panelists, there seemed to be no harbor life worth preserving for its own sake.[128]

The discussions at the conference about the harbor's and especially about its islands' future were driven by the desire to make the harbor space—just like the urban space before it in the course of urban renewal—useful, productive, and orderly. Every use, whether residential, commercial, or recreational, was to be assigned its own wedge of this detached urban enclave. So was the inner harbor north of the imaginary line between Deer and Castle Islands understood as inherently urban, devoted completely to residential and commercial development, its filling and dredging unregulated by state agencies.[129] Its status as a natural space that, as years passed, became less and less apparent at first sight was now denied altogether. As for the part south of the Deer-Castle line, where the islands grew greener and the industrial and residential development on the harbor shores more scattered and that therefore corresponded more with a conventional idea of "nature," its best possible use seemed to be recreation. Recreational development, understood as a way of preserving the islands' natural appearance, was reserved only for the area farther from Boston. For the inner harbor space, it was not even considered as an option. As Frank Gregg of the New England Basins Planning Commission put it quite unabashedly during one of the workshops: "The outer islands there is a chance to save. The inner islands . . . I wouldn't be the least bit sick to see the whole area filled."[130]

Even as the city's dignitaries debated whether to join them to the mainland or turn them into a recreational paradise, the islands of Boston Harbor sat at their comfortable distance from the city's shores, quiet and all but unaffected by the grand plans being made for their future. Despite having been generally construed as degraded and in need of "saving," little in the islands' actual appearance suggested blight and destruction. Even the 1969 documentary, which was shown at the conference, did not depict the islands as uniformly drab. Apart from the garbage heaps and ruins at Spectacle, it also presented the viewers with shots of Thompson Island and declared it to be an example of "one of the better managed and conserved areas."[131] The island was shown to be a natural-looking landscape, with a pebble beach, isolated stands of trees, and dispersed farm buildings that gave it the aspect of a peaceful rural landscape. On a closer look, the other islands did not live up to their grim reputation either. An unofficial 1978 survey of harbor-islands vegetation found more than 230 plant varieties on the islands, both native and introduced European species, including remnant stands of domesticated plants, survivors from the islands' farming past.[132] By this time, most of the islands were in fact overgrown with shrubs and trees that invaded the areas previously abandoned by farming and institutional uses.[133] Even those who ventured to Spectacle Island, by then the epitome of the "abuse" the harbor islands suffered at the hands of former generations of Bostonians, did not find it quite the desolation they expected. Alan Mikal, the author of a 1973 Boston Harbor guidebook, found Spectacle a place of surprising natural beauty, where man-made objects mixed with vegetation slowly covering the former landfill to produce a curious, unexpected loveliness. "The island," he mused about the harbor's supposedly most forlorn spot,

> could well be named "Isle of Beauty." . . . Wild rose bushes swarming with locust-sized queen bees, and fleecy-white discs of Queen Anne's Lace dot its almost impenetrable slopes. Head-high shrubs squeeze wildly among the grasses and underbrush forming an entanglement of roots and vines too thick to walk through. . . . A musical sound, similar to that made by chimes dangling in the wind, can be heard all along the beach. It is caused by waves from passing motorboats that wash the beach-glass gently back and forth upon the rocks, creating a high-pitched, symphonic echo.[134]

In August 1975, some of the harbor conference's key figures met again, this time actually out on the water. Senators Joe Moakley and Edward "Ted" Kennedy, two of the most prominent proponents of recreation in the harbor, along with the newly elected governor, Michael Dukakis, headed out to Georges Island to officially open the new Boston Harbor Islands State Park. The park, founded by 1970 legislation, had been growing slowly over the past several years, with state agencies acquiring the harbor islands, or parts of them, in a piecemeal fashion for public use. Now the islands finally promised to become the long-envisioned recreational hub for Boston. There was, however, no trace of the sprawling construction and ambitious development proposed on the islands years earlier. The alterations made to Georges, Lovells, and Gallops Islands, the core of the newly established park, were decidedly low-key and aiming at more innocuous ways of enjoying the islands: amusement parks, hotels, and apartment complexes were replaced by modest alterations made to accommodate boating, swimming, and nature walks.[135] This was a far cry from the bombastic development envisioned only a decade earlier. The minimal development the islands were experiencing was, however, fully in accord with the new understanding of the islands that began gaining ground around 1970 and that should prove to be another turning point in the city's treatment of its harbor islands.

4

THE ROMANCE OF BOSTON BAY

Reimagining the Boston Harbor Islands

Every summer Sunday in the 1960s and 1970s, a familiar sight presented itself to those walking along Boston's still somewhat rundown waterfront. At ten o'clock in the morning, a boat of the Bay State Cruises company would maneuver out of its berth at Long Wharf, turn around, and begin its way toward Georges Island, some seven miles down the harbor. On board would be several dozen passengers and their guide, a large, white-haired man with a characteristic sonorous voice, known to many in his audience from his regular radio broadcasts. He was Edward Rowe Snow, a household name to Massachusetts audiences since the 1940s when he started publishing popular books on the maritime history of New England. Now, during the short journey to Georges Island, he would alert the day-trippers to points of interest they were passing on their way across the harbor. Once the boat had rounded Gallops Island and landed at the pier at Georges, Snow would lead the group on a tour of the island's chief attraction: the century-old Fort Warren, a Civil War–era fortress that had served as a prison for captured Confederate soldiers. After Snow had guided the visitors through the fort's dark passages and tunnels, the group would eventually gather around him as he related the "most famous and dramatic" ghost story of the Boston Harbor Islands: the tale of the Lady in Black.[1]

Over the years, Snow would tell the story with varying embellishments, but the core remained the same: During the early days of the Civil War, a young Confederate soldier was captured and sent

FIGURE 22. Edward Rowe Snow, Boston Harbor's master storyteller, led boat tours of the harbor every Sunday morning during the 1960s and 1970s.— *Courtesy of the University Archives and Special Collections Department, Joseph P. Healey Library, University of Massachusetts, Boston: Collection SC-0175.*

to Fort Warren as a prisoner. When his wife heard of his plight, she donned men's clothes, traveled north, and on an (invariably) stormy night landed on the island. With the help of her husband and his fellow prisoners, she was then pulled up into the fort on a rope. No sooner had the young couple been reunited than they began plotting their escape: together with the other prisoners, they decided to dig a tunnel out of the fort. After several weeks of vigorous digging, however, the plan was discovered by the fort's garrison, and in the ensuing commotion the husband was fatally shot. The wife, now a widow, was sentenced to be hanged as a spy. Her last wish was to die in women's clothes: the request was granted, and she went to her death wearing some black robes, the only item of women's clothing that could be found at the fort. Since then, the fort was said to be haunted by a ghost of the Lady in Black: a spectral figure wearing a long black dress that would appear in the fort's dungeons and on the battlements at various times through the years, startling generations of soldiers stationed at Fort Warren.[2] The end of Snow's renditions of the story at Georges would always be the same: at a prearranged signal, an associate of Snow's would jump out from behind a dark corner, preferably with a bloodcurdling scream. Snow claimed that this ghostly reenactment could be traced back to a prank the soldiers at the fort had practiced and that "visitors who have been fortunate enough to see this performance are never likely to forget it."[3]

The tale of the Lady in Black, and the act that accompanied it, has proved a general favorite with visitors and has become as readily associated with Georges Island as Fort Warren itself. Since the time Edward Rowe Snow led his tours through the fort's corridors, the story has been perpetuated in countless retellings. In the past forty years, there has hardly been a popular book, a harbor-islands park booklet, or a newspaper article about Georges Island that has not mentioned its resident ghost story. The legend has become a standard part of ranger-led tours of the island and a fixture of a peculiar subgenre of books that occupy themselves with supernatural activity in the Boston area, with telling titles such as *Haunted Boston Harbor* or *Ghosts of Boston Town*.[4] Snow claimed that the legend of the Lady in Black "has been whispered at Fort Warren for many, many years"; however, it seems not unlikely that the story was in fact his own invention.[5] None of the numerous nineteenth-century harbor histories and guidebooks knew it; it appeared only in Snow's own lectures and book publications from the 1930s and 1940s.[6] Even if Snow did not make up his tragic cloak-and-dagger adventure entirely, he was instrumental in its ever-growing popularity. The questionable authenticity of the tale's folksy credentials did no harm to its ascension as one of the harbor's best-known and best-loved stories.

The question whether the legend of the Lady in Black was genuine island lore or merely a fabrication of Boston Harbor's greatest storyteller, however, is less intriguing than another issue it raises: why Georges Island would have needed a ghost story in the first place. One of the smaller harbor islands of about thirty-five acres of upland, Georges, like many of the islands, had in colonial times passed through the hands of several private owners from Hull and Boston who used it for harvesting wood and hay and for grazing their cattle. In 1825 its then proprietor sold the island to the city of Boston, which then passed it on to the federal government. The government proceeded between 1833 and 1861 to build Fort Warren on the island, one of several fortifications in the harbor.[7] As with the other islands used in a similar fashion, Georges Island's military career came to a close after the Second World War when the fort was decommissioned and the island itself sold at an auction in 1957 and taken over by the state's Metropolitan District Commission.[8] Throughout the city's existence, Georges Island was considered a perfectly mundane place, a site of a facility fulfilling a certain function. In the middle third of the twentieth century, however, a new way of looking at the harbor islands gradually emerged: one that emphasized their differences from rather than their connections to the city. Following the demise of some of the older island uses and filling the vacuum left by the decline of Boston's port, Edward Rowe Snow and a steadily growing number of like-minded harbor boosters set about casting the harbor islands in a new light: that of mystery and romance.

Wilderness in Boston Harbor

Unlike the spectral Lady in Black, the perception of Boston Harbor and its islands as mysterious, isolated, and profoundly different from the mainland was not Snow's invention. Beginning in the nineteenth century, people started to come to the sea in search of a wilderness quality that was missing from urban life.[9] The seemingly self-contained and isolated nature of islands made them look like antidotes of the world of the mainland that was becoming more dense and connected every day. The coast and coastal islands, on the other hand, offered well-to-do urban classes an opportunity to escape the city's congestion. The islands' ruggedness and variety allowed urban people to contemplate and enjoy a scenery that could be both picturesque and sublime in its aesthetic quality and that presented a welcome contrast to modern cityscapes. In the mid-nineteenth century, wealthy Bostonians discovered the Isles of Shoals off the Maine and New Hampshire coast and contributed significantly to changing the islands' character from fishing outposts into tourist resorts. By the late nineteenth century, they had similarly colonized the "bared and bended arm of Massachusetts" that is Cape Cod, changing the Cape's image from a bleak and inhospitable region into that of a recreational paradise and transforming its formerly lumber- and fishing-based economy into one dependent almost entirely on tourism.[10]

Wealthy Bostonians sought respite not only at these relatively remote spots but also much closer to home, on the Boston Harbor Islands. In doing so, they drew a sharp line between the islands of the inner harbor that were considered too close to the city, too urbanized, and therefore lacking in qualities that vacationing city dwellers looked for in an island, and the outer islands, particularly the Brewsters. Only these were seen as essentially island-like: isolated, weather-beaten places evocative of the land's imagined past in which the Puritan forefathers won their living through taming the New England wilderness. "The loneliness of these outer islands," wrote a *Boston Daily Globe* author in 1900, "their exposure to the full force of the ocean—which can be very ugly in this neighborhood—and their ruggedness, resulting from this unsheltered exposure, make one think that Charles Kingsley spoke truly when he said of New England: 'An iron land which only iron people could have settled in.'" [11] The islands, according to this view, were places of peril exposed to the elements and possessed a certain kind of wild, rugged beauty. They gave the impression of not having changed much since the *Arbella* sailed past them in 1630 and allowed the descendants of the Puritans to imagine them as the quintessential New England landscape through which they could rediscover their roots and reconnect with their land's past.

Late-nineteenth-century authors portrayed the more remote of the Boston Harbor Islands as places somewhat aside from contemporary civilization, quaint and exotic. An Azorean fishing colony that existed first on Long Island and after 1887 on Peddocks was a fixture of every list of colorful harbor sights, its "quaint little cluster of huts" and "olive-skinned Portuguese fishermen" who "reproduced on this faraway sister of Fayal and San Miguel the customs and sports of their homeland" adding flavor to the harbor's perceived otherness.[12] The harbor islands were presented as picturesque in form and if peopled at all then only sparsely and with maverick characters who added to rather than detracted from the islands' romantic charm. The *Globe* reporters who in 1900 made a trip to the outer islands described in great detail their encounter with one of these curious islanders, the lobsterman James Turner, the "emperor of Calf Island." Turner, who had by then lived on the island for forty-six years, "having for company only his wife and his Newfoundland dog Prince," was the son of another longtime islander who became notorious for scaring off, with a shotgun in hand, Boston boating parties that made the mistake of landing on his island.[13] Grape Island in Hingham Bay was home to another harbor hermit, generally known as Captain Smith, who, according to a popular account, was just as likely to treat visitors to stories of his youth on board a slave ship as to greet them with shots from his musket upon arrival.[14] While such behavior would have been entirely out of place on the mainland, it seemed only fitting in the small, isolated world of the harbor archipelago. The islands seemed to exist in a slightly different reality than that of the mainland: a more adventurous and exotic one, attracting characters who seemed as fallen out of time as the islands themselves.

To go to the islands was to escape the strains of daily life. When in July 1891 Helen Augusta Whittier, an entrepreneur and art teacher from Lowell, and her three female companions set out for Great Brewster to spend nearly two weeks there, they regarded it as an extraordinary adventure that allowed them to leave everyday reality behind, even to the extent of temporarily shedding their identities. In the diary Whittier kept about their time on the "enchanted isle," she refers to herself and her fellow adventurers as the "Autocrat," the "gentle Aristocrat," the "artistic Acrobat," and the "veracious Scribe," but never by their names. The island provided them with an opportunity to embrace, if only temporarily, a leisurely lifestyle of days filled with reading, sketching, taking photographs, going on walks around the island, and having moonlit clam and lobster dinners. When "Ye Merrie Trippers," as Whittier referred to herself and her companions, were leaving Great Brewster at the end of July, the last entry in their island journal was permeated with a sense of having experienced an extraordinary time, made possible by their trip's isolated island setting: "At 10:30 a.m. we said goodbye to our enchanted isle, so difficult to reach, so hard to leave. . . . We leave behind us the uneventful, idyllic days, like no others in our lives, with their placid serenity, their pleasant spice of labor, the unruffled happiness of accustomed comradeship and all the glory of the sea and sky."[15] The sense of isolation that Great Brewster offered seemed to make existence slower and simpler, as if the water expanse that separated it from the mainland also signified an interval between different ways of life.

Other notable Bostonians of the late nineteenth century were determined to secure this transient idyll for themselves in a more permanent manner. Benjamin Pierce Cheney Jr., the director of Wells Fargo & Company, and his wife, the actress Julia Arthur, owned a home on Calf Island, while Augustus Russ, a wealthy lawyer and one of the founders of the Boston Yacht Club, owned the whole of Middle Brewster and had a summer "cottage" built on the island.[16] The association of islands with summer leisure did not die down with the close of the nineteenth century: in 1967 the Andover stockbroker Jim Fox bought Green Island as a vacation home for his family of nine.[17] Similarly, Andrew P. Quigley, the publisher of the *Winthrop Sun-Transcript*, was the proprietor of a portion of Middle Brewster, and he and his family used the island for picnics and camping.[18]

While sharing the view of islands and the coast as wilderness worthy of preservation, the late-nineteenth-century park movement was opposed to the notion that the harbor islands should remain the playground of a privileged few. When the state of Massachusetts

FIGURE 23. A watercolor of the "enchanted isle" of Great Brewster from Helen Augusta Whittier's 1891 album. To Whittier and her companions, their time on the island represented an escape from the ordinariness of life on the mainland. *—Courtesy of Schlesinger Library, Radcliffe Institute for Advanced Study, Harvard University.*

established the Metropolitan Park Commission in 1892, one of the first public bodies in the United States tasked with the planning of a comprehensive park system on a regional scale, its searching gaze soon fell on the harbor islands.[19] The commission's secretary was Charles Eliot, a young Boston-based practitioner of the new profession of landscape architecture, who became a key figure in the early development of Boston's metropolitan park system. Eliot argued that the surviving enclaves of natural beauty around Boston needed active preservation before they became engulfed and destroyed by urban expansion.[20] Looking for large swaths of natural spaces that would provide a welcome contrast to the "noisy ugliness of towns," Eliot remarked in a report to the commission in 1893:

> We have found that the metropolitan district of Boston lies, even at this late day, between two wildernesses; on the one hand the untamed heights of the rock-hills, on the other the untamable sea. . . . Thus has nature placed and preserved at the very gates of Boston riches of scenery such as Chicago, or Denver, or many another American city would give millions to create, if it were possible. Stupid indeed will be the people of greater Boston if they fail to perceive and attend to their interests in this matter before the opportunity is lost.[21]

In his view, Boston Harbor and its shore constituted a key element of Boston's system of open spaces. Eliot's proposals came at a time when the seaport's primacy over the harbor space was virtually unchallenged: when suggesting that portions of it should be devoted to recreation, he was careful to point out that recreational use would by no means obstruct shipping and would in fact use only those corners of the harbor where maritime development was not an option. The shore of Quincy Bay, he argued, would make an equally excellent seashore reservation as the already existing, and widely popular, Revere Beach, "because the shallowness of its water similarly precludes commerce." The public acquisition of as large a portion of the seashore as shipping interests would permit was in Eliot's eyes just as important for the success of the metropolitan park system as its two mainland cornerstones, the large reservations in Blue Hills and the Middlesex Fells.[22]

Creating a natural public playground replete with grand ocean vistas and sea breezes was not the only objective Charles Eliot and the Metropolitan Park Commission had in mind when they sought to incorporate the harbor islands into the city's park system. By preserving and "improving" patches of forests, hills, and hitherto undeveloped river and sea shores, they aimed to accomplish a goal that historian Michael Rawson has called "recreating the Puritan wilderness." Naming and shaping the landscapes in certain ways was meant to be evocative of the shape of the land that their Pilgrim forefathers had encountered when they first came to New England.[23]

Re-creating allegedly primal forests on the harbor islands was also a way of making the harbor more pleasing to the eye, thus facilitating at least a passive enjoyment of the coastal landscape. Many of the harbor islands were not available for active recreation, being used either by the federal government or by the city of Boston for institutions such as forts, hospitals, or poorhouses, but they could still form at least a virtual part of the string of natural places encircling the Boston metropolitan area. With only a small effort, Eliot suggested, the harbor islands "might easily be clothed in foliage, to the great improvement of the scenery of the bay."[24]

Eliot's proposal echoed the statements that his mentor and business partner, Frederick Law Olmsted, made to the Boston Park Department only five years earlier. In his report, submitted to the Park Department in December 1887, America's foremost landscape architect praised Boston Harbor as one of the city's most attractive features, yet at the same time called for improvements in the harbor's physical appearance, especially as far as the "hard-featured, bare, bleak, and inhospitable aspect of the headlands and islands" was concerned. In his criticism, he linked the harbor islands' present aesthetic deficiencies to the loss of their "original beauty" that they had presumably possessed in John Winthrop's time:

> Stumps, that still remain upon the most exposed, the rockiest, and bleakest of the islands show that they formerly were wooded. Once cleared, a second growth has been prevented by cropping and pasturing. . . . Hence the scenery of the harbor has been and is every year being despoiled more and more of its original beauty; its artificial features are becoming more and more disagreeably conspicuous relatively to its natural features, and in these respects it is becoming less and less attractive.[25]

Reforesting the islands, Olmsted asserted, would not only bring back welcome reminiscences of the "Puritan wilderness" past and emphasize the islands' historicity but also contribute significantly to their present-day aesthetic merit. "Almost at once," he assured the Park Department officials, "the sapling plantations will give a pleasing softness and geniality to those elements of the scenery that are not contributive to its picturesque ruggedness. Three years after the planting is finished the harbor, as a whole, will have acquired a decidedly more good-natured, cheerful, and inviting character."[26] Planting the islands would thus have both evoked their historicity and set them off visually from the cityscape, creating the impression of a wilderness, while at the same time assigning them a place in Boston's system of open spaces. Visually detached, they would have been drawn into the "emerald necklace" of the city's recreational areas.

Unlike Blue Hills and portions of the harbor shore, the Metropolitan Park Commission did not manage to acquire the harbor islands as parkland at the turn of the century. The idea of making them aesthetically pleasing by giving them a cover of foliage, however, kept resurfacing from time to time, although it had temporarily lost its heritage spin. In the early years of the twentieth century, it was taken up by the City Beautiful movement, represented in Boston by the Boston 1915 campaign, a program devoted to coordinating the efforts of various educational, recreational, health, and charitable organizations toward creating a progressive, healthful, and beautiful city.[27] These groups decried the "barren nakedness" of the harbor islands and invited the public to imagine how, by planting groves of trees, the islands may be transformed into "beauty spots" instead.[28] In the 1930s, Boston City Hall proposed to combine the face-lift of the harbor islands with the implementation of New Deal measures. According to a suggestion made by Mayor James Curley, the islands would be reforested by the Civilian Conservation Corps, a recently introduced public work-relief program.[29] Abandoning any plans of incorporating the harbor islands into the city's park system, this and other similar proposals of the era focused solely on beautifying the islands and making them the visual pivot points of the harbor. As Curley declared, "Trees planted on the islands of Boston Harbor would provide the finishing touch for making this harbor the most beautiful in the country."[30] Other proposals went even further: in 1934 the prominent landscape architect Warren H. Manning recommended planting the islands with exotic flowers and shrubs in order to make them into a grand gateway to Boston or even into sights that visitors would come specifically to see. Such makeover, Manning asserted, which would have included planting the islands with honey locusts, lilacs, weeping forsythias, Osage oranges, and tamarisks, would "attract as many visitors to Boston's floral islands as visit Washington for the cherry blossom display."[31] Despite outraging the New England Nurserymen's Association, which adopted a resolution saying that planting with exotics and the "startling and showy floral effects proposed" would "most seriously and permanently detract from the far-famed beauty of our natural New England landscape," the plan spoke directly to the perception of the Boston Harbor Islands as inherently different from the city, yet a part of it in a certain way. Set off aesthetically from the city—whether by planting chestnuts or *Genista scoparia*—they were at the same time the city's front yard, their beautification ultimately the city's beautification.[32]

To landscape architects like Manning or Eliot, the harbor islands were primarily places with a recreational potential: a useful suburban wilderness, made valuable by its closeness to the city. To them, the harbor and its islands were a perhaps mismanaged but unquestionable part of the urban orbit. Theirs was a time when Boston Harbor, despite the port's slowly diminishing standing, was still one of the centerpieces of the city's economic identity. In the 1890s when Charles Eliot drafted plans to include the harbor islands in the metropolitan system of nature preserves, and even in the 1930s when Warren Manning proposed to turn them into Boston's flower garden, the harbor islands were places well known to many Bostonians. They were the sites of institutions and businesses some of which it was hard to ignore, like the garbage-reduction plant on Spectacle Island. Some of them were inhabited by small populations of workers and fishermen: a 1937 survey by the Harbor Police found 153 adults over twenty years of age living on the islands.[33] Despite several small recreational capsules along the harbor shore and on the outer islands, the primary way in which people living around Boston Harbor related to the harbor space was through work: at the city's many wharves, in its shipyards, on its fleet of fishing boats. For these people, the harbor held little romance: for them, it was the site of everyday labor, a space they experienced firsthand on a daily basis. Its characteristics—its tides, flats, currents, and channels—were all familiar and devoid of any sense of mystery, and the harbor islands were part of this space.

However, the early-twentieth-century decline of Boston shipping and the mid-century changes in port technology toward automation as well as the decline of Boston's fishing industry meant that the number of this workforce of the sea dwindled and with it the number of people for whom the harbor was as commonplace as the field or the factory floor. The understanding of the harbor and of the harbor islands as a working landscape was slowly disappearing, creating a vacuum not only of use but also of meaning. The old harbor uses' gradual demise made the harbor space open to new interpretations and allowed Bostonians to imagine their harbor in new ways. One of them was the 1960s city planners' attempt to claim the harbor and its islands as real estate and Boston's extension. Ultimately, however, another narrative about the harbor proved more successful and durable: the one told by Edward Rowe Snow.

The Romance of Boston Bay

Born in 1902 in Winthrop, Edward Rowe Snow reportedly had some experience with the life at sea himself: after finishing high school, he worked on oil tankers and sailing ships for several years. Even in his later years, when he came to be firmly based on dry land, he spent much time on or over the water, but this time engaging in leisure activities: canoeing with his wife, Anna-Myrle, and their daughter, Dorothy, among the many islands of Boston Harbor and acting as the "Flying Santa," delivering Christmas gifts in a small plane to lighthouse keepers and inhabitants of islands along the New England coast.[34] After several years of working as a high school teacher of history in Winthrop, he quit his job to devote himself full-time to writing and lecturing about his lifelong passion: the history of the Boston Harbor Islands and the Atlantic Coast.[35]

From 1935, when his first book came out, until his death in 1982, Snow published nearly a hundred volumes, in most of them navigating the coastal waters of New England and the Atlantic Seaboard. In addition to his tours of the harbor and Georges Island, which, though popular, reached only a limited number of people, he gave public lectures on harbor topics and, through the 1940s and 1950s, hosted a radio show called *Six Bells* that brought maritime lore to children and teenagers. Ships, islands, and tales of the sea enlivened his newspaper columns in Quincy's *Patriot Ledger* and in the *Boston Herald* and were also the topic of his television appearances.[36] Snow became a prominent personality of the region's public life, and for generations of readers, viewers, and listeners in the Boston area who were even remotely interested in the Boston Harbor Islands, he became the most authoritative voice on the subject.

Edward Rowe Snow's understanding of islands was deeply rooted in the romantic artistic imagination of islands as remote, isolated "locales of desire." The sea to him was a landscape of drama, the "sublime theater of crisis and catastrophe" of Byron's and Coleridge's poems and Turner's paintings.[37] The way Snow described the islands along the North Atlantic coast can be traced back to the influence of island fiction such as Jules Verne's *The Mysterious Island*, Jonathan Swift's *Gulliver's Travels*, or Robert Louis Stevenson's *Treasure Island* that he evoked himself: "In all of literature, what stories rouse one's sense of adventure like *Treasure Island* or *Robinson Crusoe*?"[38] The author Edward Rowe Snow, however, was no Daniel Defoe: his books rarely transgressed the genre of trivial literature and consisted mainly of series of anecdotes featuring the classic props of maritime tales: pirates, shipwrecks, and chests of gold. He portrayed the waters of the North Atlantic coast as a world apart where the dreary reality of the mainland held no sway: a wonderland swarming with sea serpents and giant squids with "tentacles over thirty feet long."[39] Like the sea around them, the islands were places of wonder. Their remoteness was not a mere geographical fact: translated as isolation, it was the foundation of their uniqueness. An island, Snow insisted, was "a body of land surrounded by adventure, romance, mystery and excitement."[40] Little of the actual hard work, dreariness, or repetitiveness of life at sea had a place in Snow's books. Furnished with sensational titles such as *Marine Mysteries and Dramatic Disasters of New England* and *Ghosts, Gales and Gold* as well as *Piracy, Mutiny, and Murder*, they told the "history" of the waters and islands off the Atlantic Coast as a series of thrilling stories typically involving shipwrecks, doomed romances, miraculous rescues, ghosts, pirates, and buried treasure.[41] In Snow's hands, the entire Atlantic Coast transformed into a realm of the strangest occurrences where the ghosts of the drowned walked the earth, monsters of the deep threatened, and mysterious shipwrecks beckoned. While the world of the mainland, and especially of the city, was governed by "hectic rush and dull routine," islands were home to extraordinary events and adventure.[42] The water acted as a boundary between the dreary present and an exciting, if unspecified, past. Tales of the islands' thrilling history framed them as places outside the ordinary, everyday existence. In this imagination, the tidemark signified a boundary between worlds: here toil and drudgery, there adventure and romance.

Edward Rowe Snow was not the only one to imagine islands in this way: the perceived remoteness and romantic ruggedness of places such as Mount Desert Island in Maine or the Isles of Shoals, or even the outer Boston Harbor Islands, had been exploited by generations of city dwellers eager to escape the bustle of urban life.[43] Snow, however, sought to expand this interpretation to all islands of Boston Harbor. In his first and arguably most popular book, *The*

Islands of Boston Harbor, which first came out in 1935, he painted a picture of an island group that abounded in all the qualities of mystery and adventure that more remote and more pronouncedly insular places possessed, even if the Boston Harbor Islands' very ordinariness and their mundane histories resisted his attempts at romanticizing them.

By the 1930s, the Boston Harbor Islands were by most standards a decidedly urban archipelago. For centuries, they had been extensions of Boston, their uses governed by the city's needs and their flora, fauna, and very shape determined by the city's changing demands. Edward Rowe Snow, however, presented them to the readers of his books and the audiences of his lectures as a fascinating landscape of legends and thrilling adventures, as their city's very own treasure islands. "Probably no comparable area in the entire United States has been the scene of so many interesting and exciting events as the islands of Boston Bay," he wrote in 1950. Snow not only bestowed on Boston Harbor its ghost story of the Lady in Black but also found scores of "secrets," "gruesome discoveries," "thrilling adventures," or at least "strange occurrences" for each and every one of the islands.[44] The eroded island of Nixes Mate provided the setting for a genuine pirate story. Little Brewster with Boston Light was the stage of fearsome storms, miraculous escapes, and tragic shipwrecks. Castle Island supplied a dramatic shooting duel, a passing sea serpent, and a skeleton in the dungeon.[45] There was even an actual treasure island: In 1782 the French warship *Magnifique* ran aground near Lovells Island. In 1920, as Snow reported, "the lighthouse keeper, Charles H. Jennings, was digging in his back yard when he came across several blackened discs. He scrubbed them vigorously to remove the dirt and rust and found that they were gold pieces, obviously from the *Magnifique*. When he went to Boston, he sold several of them for $29 each."[46] Although the story clearly employed the well-worn fairy-tale device of an unassuming yet deserving person coming to indescribable riches, the particulars of the story happening on a specific date to a specific man, as well as the rather practical detail of the exact worth of the coins, made the event seem real and full of promise. The islands were clearly—in contrast to the all-too-familiar and commonplace milieu of the city—places outside everyday experience, places where it was possible to find a buried treasure. If you go to the islands, the implication seemed to be, marvelous things may happen to you, too.

When rendering the Boston Harbor Islands as places of romance and mystery, Spectacle Island proved particularly challenging. Turning the site of a landfill and a notorious nuisance into an enchanted isle required considerable power of imagination. Snow possessed it: Although the island lacked ghosts, pirates, and romantic ruins, it was home to José Safarino, also known as Portuguese Joe, who lived in a cottage on the eastern side of the southern bluff. When Snow visited him there, Safarino spun story after story of harbor lore, filling the role of the wise old man whose memories connect the present to another, simpler, and purer time of the past. Other than the "island fisherman," Spectacle had admittedly not much to offer in the way of excitement: the only danger that lurked on the island was the garbage dump, where, as Snow warned his readers, "the rats grew as large as cats." While devoting considerable space to the island's colonial history, the garbage factory that at the time of Snow's visit was the island's most conspicuous feature received only a passing mention.[47] Snow's island world was defined by the romanticized past, not the ordinariness of the present.

Snow's narratives helped suppress the harbor islands' historical connections to the city, but, interestingly, they also helped preserve some of the historical buildings on the islands, especially those that were the stages of some of his stories. When rumors surfaced in the late 1950s that Georges Island, which was about to be sold by the General Services Administration at an auction, was being contemplated as a landfill or even a storage site for radioactive material, Snow became the president of the Society for the Preservation of Fort Warren, formed in response to these plans.[48] The fort's presence on the island and the legend associated with it played a major role in the deliberations of what was to happen with the island. Newspaper reporting on the auction emphasized that

Georges was a "historic island," "replete with military legend," and that it would be far more fitting to turn it into a "national shrine" than to forfeit it to "prosaic business use."[49] The Lady in Black legend was also mentioned. In the end, the island's historicity proved more highly valued by the state than its potential use as a storage site: the Metropolitan District Commission took Georges Island over from the private bidder, and in 1961 Fort Warren was opened to the public.[50] The Lady in Black and the boost her legend meant for the island's popularity thus proved instrumental in the actual preservation of the fort.

In the five decades of his career as a writer and harbor-islands advocate, Edward Rowe Snow achieved a striking level of success in the Boston area. His lectures enjoyed wide popularity: even at the start of his career in the 1930s, a talk about gold and ghosts in Boston Harbor that he gave at the historic Old State House to the Bostonian Society "drew the largest attendance the oldest members could remember ever having seen there."[51] His *Islands of Boston Harbor* was published in three editions over the years, while his articles on the subject in Boston-area newspapers went into the hundreds.[52] His vision of the Boston Harbor Islands as extraordinary places and embodiments of romantic insularity was, though, not entirely unique even in the 1930s. In 1932 the Boston postmaster Patrick J. Connelly published a slim volume about the harbor islands with the poetic subtitle "Green Isles of Romance." In the introduction, he, too, staged Boston's hitherto relatively inconspicuous islands as places of natural beauty and witnesses to the "romance of the ages." He wrote, much in the vein popularized a little later by Edward Rowe Snow: "Through all the centuries, men in all climates and lands have been building romantic visions of the beautiful green isles far off, where their dreams would come true. . . . And to those who first sailed into and up the Harbor of Boston, passing the green isles and seeing the beautiful contour of their lines, all the dreams and pictures of their minds, visions and hopes unfolded into a reality."[53] Connelly, however, did not possess Snow's storytelling talent and his showmanship; while the literary postmaster slipped back into obscurity, Snow, whose *Islands of*

Boston Harbor: Their History and Romance appeared only three years later, rose to prominence, and his narrative about the Boston Harbor Islands became canonical.

Snow's way of romanticizing the world of the harbor islands tapped into a rich vein of escapism among the Boston urbanites. The islands of Edward Rowe Snow were as mysterious, romantic, and full of adventure as everyday life in the city was pedestrian, prosaic, and dull. The harbor islands with their colorful legends offered an easy escape from all this to anyone who could afford a boat ticket. This version of adventure was a tame one: all the deadly disasters, daring deeds, and dreadful discoveries happened in a past long gone, and they could be experienced by the visitors and readers quite safely through retellings and reenactments, sending pleasant shivers down their spines. In this sense, the islands of Snow's books, lectures, and broadcasts existed somewhere on the border between a thrilling land of adventure and a theme park. By undertaking a journey to the islands or—to a lesser extent—by simply attending one of Snow's lectures, Bostonians could access the islands' allegedly romantic past. By visiting the fort where the Lady in Black used to walk the battlements, the island where the pirate's skeleton was found, or the lighthouse that had weathered countless frightful storms, they could take part in the harbor lore. Taking a boat to the islands became a way of walking through the time tunnel into a more romantic, more exciting, and more eventful world of the islands' imagined past.

Edward Rowe Snow's approach to the harbor islands was, most significantly, a sharp departure from how the islands had been historically seen from Boston's waterfront. Earlier generations of Bostonians mostly looked at the islands as useful adjuncts of the city, made valuable only by the services they could provide to meet the city's current needs. Whether it was for resources such as timber, hay, or building stone, their physical isolation and ability to absorb unsavory businesses and unpopular institutions, or their vacant space ready to solve Boston's urban problems, the islands' value almost always derived from the uses the city had for them. Snow's ideas were radically different: although he also played the card of the islands as refuges of calm for the city-weary population, he seemed to argue that the harbor islands possessed an inherent value that rested on their "history and romance." The harbor islands, his books insisted, were places in their own right, with their own interesting histories that were in many cases independent of what was happening in Boston. His harbor islands were storied places, and even if some of the stories he attributed to them were not strictly speaking true, they still brought the islands out of Boston's long shadow and made them shine as marvelous, exciting places.

Edward Rowe Snow's vigorous activity helped raise awareness of the harbor islands among the population of Greater Boston. Perhaps most important for the future of the islands, he influenced and inspired a new generation of Boston Harbor boosters: mostly young people who shared his fascination with the world of the "green isles of romance" and were equally concerned about their preservation. Suzanne Gall Marsh, the founder of the Friends of the Boston Harbor Islands, the largest volunteer organization devoted to the preservation and promotion of the harbor islands, has cited meeting Snow as a major inspiration that led her to start a harbor-islands volunteer program in 1979.[54] Similarly, Jeremy D'Entremont, a lighthouse enthusiast, popular historian, and author of more than twenty books about the lighthouses of New England, has credited Snow's books with sparking his lifelong interest in maritime lore.[55] Snow's work found reception even in the highest echelons of Massachusetts politics: Governor Francis Sargent wrote an enthusiastic review about the second edition of Snow's *Islands of Boston Harbor* in which he lamented that "these islands, steeped in historical tradition, have been allowed to deteriorate and decay" and welcomed Snow's book as a contribution to renewing interest in the "rich and vast heritage of Boston Harbor."[56] Edward Rowe Snow's vision of the Boston Harbor Islands as special places that differed fundamentally from the city and were worthy of protection proved contagious. For the 1970s generation of harbor-islands champions, however, the harbor was not solely a space defined by legends and historical anecdotes; they were the first generation to see Boston Harbor as a valuable and endangered ecosystem.

FIGURE 24. Edward Rowe Snow (*third from the left*) with members of the Boston Harbor Islands Volunteer Corps in 1981. Founder Suzanne Gall Marsh is on the right. Snow's books and public lectures served as inspiration for the next generation of harbor boosters.—*Courtesy of the University Archives and Special Collections Department, Joseph P. Healey Library, University of Massachusetts, Boston: Boston Harbor Islands Mass. Memories Road Show collection.*

5

A NEW SPECTACLE

Conservation and Recreation in Boston Harbor

In the summer of 1979, two young women met on Gallops Island in the middle of Boston Harbor. Suzanne Gall, at twenty-nine the older of the two, came to Boston from Bridgeport, Connecticut, several years prior. An outdoors enthusiast, she had spent the previous summer exploring her new city's harbor islands; she had also met Edward Rowe Snow and became captivated by his islands stories. Her companion, Katherine "Kathy" Abbott, was a twenty-two-year-old college student of landscape architecture and regional planning who grew up at Cape Cod. After interning for various government branches, that year she went looking for a summer job with a state agency and landed with the Department of Environmental Management (DEM), which ran the Boston Harbor Islands State Park. Now they both found themselves out in the harbor as members of a brand-new harbor species: the island manager. After the state park started operating in the early 1970s, the DEM introduced an island-manager program in 1976, peopling the islands with seasonal resident staff. The young rangers would live on "their" island through the summer; they would assist visitors, introduce them to the history and nature of the islands, and also act as law enforcement. Not least, they would help popularize the understanding of the harbor islands as places where nature and history met and as public open spaces for Boston and surrounding communities.[1]

As with the other harbor islands, Gallops Island's identity as a recreational spot was still very new. Its history followed the typical

Boston Harbor trajectory: used for farming in the seventeenth and eighteenth centuries, it became the site of Boston's quarantine hospital in the 1860s and of an army radio school during the Second World War. Since then, the island was "deserted," ruins of buildings slowly overgrowing with sumac, and stands of mock orange, snowberry, and forsythia from the former hospital's garden spreading over the island in all directions. It had become accessible by public boat service only in 1975, and not much in its history suggested a future as nature preserve and place of leisure.[2]

The young island managers, however, had no trouble seeing the island in this light. Both Kathy Abbott and Suzanne Gall came from community-minded and environmentally conscious families, so working for the state park was for them a form of public service, introducing Bostonians to their city's new recreational opportunity. As Suzanne Gall recalls, Edward Rowe Snow was not only an inspiration but also a regular presence during their time on the island: "Every weekend as the morning *BayState* ferry boat passed Nixes Mate and Gallops Island, we would hear Mr. Snow's melodious voice wafting across the harbor." His stories that cast the harbor islands as exciting places in their own right became one of the sources of Suzanne Gall's conviction that the islands should be preserved for public enjoyment. Finally, both young women were very much children of their age: the era's environmental awareness allowed them to see Gallops Island as a resource to be valued, protected, and preserved.

The island managers' daily life on the island resembled that of any park ranger: they would greet the visitors, give them tours of the island, and clean up after them. After the last day-trippers had left in the late afternoon, the day's manager (both of them would be there on the weekend, but only one during the week) would stay behind, alone on her island. The sense of place could be overwhelming: Gallops Island seemed both like a part of a larger whole, surrounded by other islands on all sides, and like the very edge of the continent. Kathy Abbott would sometimes climb the island's bluff at sunset and, with a (warm) beer in hand, look east beyond the Brewsters to the open ocean and then west where the sun set over the Boston skyline and the three million people of the metropolitan region. The overlapping sense of isolation and belonging, the island's urban past and its natural present all combined into a transformative personal experience. For Kathy Abbott, the time on Gallops translated into a lasting realization: that of the harbor islands' enormous potential to introduce thousands and millions of people to the "delicate balance between the built environment and the natural world" and to educate them about their own impact and "the responsibility that creates for every individual."[3]

That summer on Gallops Island marked the beginning of two outstanding Boston Harbor careers. That very year, Suzanne Gall founded the Boston Harbor Islands Volunteer Corps, a volunteer program aimed at assisting the state park's island managers and acquainting the public with the harbor islands. Cosponsored at first by the Sierra Club and the Appalachian Mountain Club, the Friends of the Boston Harbor Islands, as they have been called since 1982, have become over the years the largest volunteer group in Boston Harbor, with Suzanne Gall at the helm. Kathy Abbott went on to work as a planner for the DEM, rising through the ranks to eventually become the first commissioner of the Department of Conservation and Recreation in 2003. Since 1996 she has also been involved with nonprofit organizations promoting Boston Harbor and its islands, most recently as president and chief executive officer of the harbor-wide group Boston Harbor Now.

Suzanne Gall and Kathy Abbott's stint as island managers was also a moment that heralded a new era in the perception and use of the Boston Harbor Islands. The new state park and the young women's ability to see the former quarantine island as a charming place worthy of protection and as a potential site of environmental learning were grounded in a way of looking at the islands that was both new and familiar. It combined the tradition of harbor recreation, the notion of the islands as places with their own history, and the emerging environmental sensibility into a new island narrative. The Boston Harbor Islands were yet again regarded both as the city's asset and as places apart. They have become recreational resources whose value rested on their historicity, their ecological communities, and their natural aspect—and which, in order to be enjoyed, could not be allowed to become urbanized.[4]

The Boston Harbor Islands and the Environmental Decade

The new way the Boston Harbor Islands were imagined and used owed much to the rise of environmental concerns in 1970s America. This environmental awareness and the environmental legislation it gave birth to did not emerge overnight and were grounded in a steady growth of civic involvement and legislative protection of natural resources in the postwar era. The "environmental decade," however, still represented a watershed in the prevailing perceptions of the natural environment and in the advancement of environmental causes.[5] The new environmental sensibility and the mandate it provided for agencies concerned with the state of the nation's waters, air, and soil brought about such breakthrough federal legislation as the 1969 National Environmental Policy Act, followed by the establishment of the Environmental Protection Agency (EPA) in 1970 and the significant expansion of the existing Clean Air and Clean Water Acts.[6] Massachusetts state legislation followed suit only a few years later: the starting point of the state's environmental turn was the 1972 adoption of an amendment to the Massachusetts Constitution that took an expansive view of the meaning of the term *environment* and asserted the state's citizens' right to a clean environment and the state's leading role in providing it: "The people shall have the right to clean air and water, freedom from excessive and unnecessary noise, and the natural, scenic, historic, and esthetic qualities of their environment; and the protection of the people in their right to the conservation, development and utilization of the agricultural, mineral, forest, water, air and other natural resources is hereby declared to be a public purpose."[7] Also in 1972, the Massachusetts Environmental Policy Act required state agencies to take into account the environmental impact of their decisions, while the Department of Environmental Quality Engineering and the Executive Office of Environmental Affairs (EOEA), established in 1974 and 1975, respectively, were created to oversee and enforce the compliance with the state's environmental legislation.[8]

The novelty of environmental thought could be seen clearly at the 1969 Boston Harbor Islands conference at the Massachusetts State House. Taking place a mere two weeks before the National Environmental Policy Act was passed, and seven years after the publication of Rachel Carson's *Silent Spring*, some of the conference participants showed a striking indifference to and even ignorance of ecology and environmental concerns. A conference panel assigned the task of discussing the future of Boston Harbor from the perspective of ecology and oceanology spent a considerable amount of time debating what oceanology even was. As for the notion that ocean ecology was a value to be protected, the panel's chairman, Frank Gregg of the New England Basins Planning Commission, dismissed the idea without much ceremony: "I don't think it is. It's a dismal science which is just as neutral as economics or any other. . . . But the objective is not ecology. The objective is the management of the Harbor's resources."[9] Although lone voices from the audience expressed the view that the harbor's ecology and its health had an intrinsic value that was independent of the demands the city placed on it, the prevailing opinion was that the harbor's value rested in its usable resources, whether they were fisheries or recreational potential that could bring tourist dollars to the region. Most of the panel's participants saw no point in protecting resources that could not be readily translated into monetary terms.

At the same time, the conference was attended by a number of politicians and officials who subscribed to the view that Boston Harbor and its islands had the most value for the public as an area where the preservation of an unpolluted and healthy natural environment was coupled with moderate recreational use. Charles G. Carothers, deputy assistant secretary at the federal Department of the Interior, spoke about the danger of estuarine areas across the nation being filled for residential development, a phenomenon he referred to as "estuarine habitat destruction," and reminded Boston and Massachusetts politicians that plans for land making in Boston Harbor needed to consider the impact on the harbor's twenty-three thousand annually wintering waterfowl and its twenty-one species of fish. Harbor pollution, Carothers asserted, was not only an aesthetic drawback but in the first place a grievance for beach recreation and a threat to marine life. Massachusetts governor Francis Sargent, a well-known champion of environmental causes who gave the keynote address at Boston's first Earth Day celebration in 1970, pleaded for the preservation of Boston Harbor as a recreational area in the manner of the Cape Cod National Seashore, which he had helped bring to life. The organizer of the conference, the legislature's Special Commission on the Boston Harbor Islands headed by state Senator Joe Moakley, recommended managing the harbor through a program of conservation and recreation and proposed legislation that would "preserve and enhance productive harmony within the ecological systems of the harbor."[10] Some of these views were hardly less utilitarian than assessing the harbor's value in terms of its measurable marine resources: instead of shellfish and gravel, it was recreational use that could be extracted out of the harbor and its islands. Nevertheless, the prominent place that the preservation of the islands as a natural area took up in the conference discussions was a sign of its rising importance in the contemporary public discourse.

It was, however, local and regional nonprofit groups who were the torchbearers of the environmental agenda in Boston Harbor. Since the late nineteenth century, Boston had been the center of one of the nation's most vibrant conservation-group networks. The Massachusetts Audubon Society, the Appalachian Mountain Club, the Massachusetts Forest Association, and the Trustees of Reservations were all formed in the Massachusetts capital. The period from the late 1960s through the 1980s saw the emergence of a new generation of advocacy groups: ones that combined local improvements with subscribing to the general environmental cause of the era. One of the first was Quincy's Save Our Shores, a local organization formed in the late 1960s in direct response to the Expo 76 project and the plans to fill parts of Dorchester Bay. Despite its local focus, SOS developed a strong environmental agenda, arguing for the whole Boston Harbor to be declared a national park. The Boston Harbor Associates, established in 1973, brought together Boston's business, planning, civic, and academic communities in an attempt to balance recreation and business interests in the harbor. Although sometimes criticized for its prodevelopment stance, TBHA helped launch two of the most visible and influential harbor advocacy groups: the Friends of the Boston Harbor Islands in 1979 and Save the Harbor/Save the Bay in 1986. Even organizations not exclusively dedicated to Boston Harbor felt obliged to chip in: the Sierra Club established a harbor task force in the 1970s that was involved in advocating for the harbor-islands state park. Despite their different standpoints, all these groups contributed to a heightened awareness in the surrounding population of the harbor, its recreational potential, and its environmental problems.[11]

What was a fledgling effort in the late 1960s became the standard approach within the next decade. In the public discourse of the 1970s and 1980s, the prevalent image of the harbor islands became that of an ecologically valuable natural area worthy of protection. This way of seeing the islands mingled with the already established narrative of the harbor islands as a historic landscape providing access to Boston's and Massachusetts' colorful past. Local citizen organizations like Save Our Shores called for the "restoration, beautification and preservation" of the harbor islands, praising in equal measure their "historical significance and picturesque beauty" and their importance for the "ecology of marine life."[12] The Boston Harbor Islands Volunteer Corps started in 1979 by Suzanne Gall was devoted to promoting both the islands' natural beauty and their historicity. Although sponsored by prominent conservationist organizations, the corps' successor organization, the Friends of the Boston Harbor Islands, did not adopt as its symbol one of the more rugged and natural-looking harbor islands but rather chose the striped wooden beacon on Nixes Mate—one of the harbor islands most conspicuously shaped by human activity.[13]

The islands' double quality of open space and historicity was taken up by the press, too. During the 1970s and 1980s, writers for Boston-area newspapers frequently reported about the harbor islands as places that had been mistreated in the past, but now, if properly managed, could offer the tranquillity of nature and a sense of their colorful histories to their visitors. In 1977 nature writer Ruth Aronson spent the summer as island manager on Great Brewster Island (where by that time no trace of Ye Merrie Trippers' cottage could be found) and managed to pack into her report for the *Globe* the subtle pleasure of listening to the crackling of barnacles and periwinkles on the shore at dawn, the migration of monarch butterflies, and reminiscences of the Revolutionary War.[14] The *Boston Herald* and the *Offshore Magazine* in the early 1980s featured stories about Peddocks Island that emphasized both the island's exceptional calm and the slightly eerie feeling that those walking its overgrown roads experienced. This feeling of unease was readily attributed to a seventeenth-century massacre that allegedly happened on the island and was described in great detail in Edward Rowe Snow's books. "Is it possible that the tragedy which befell the island over 300 years ago still hung in the atmosphere?"[15] Such stories made the islands into places where nature and history were closely intertwined and whose uniqueness rested precisely on this combination of perceived historicity and scenic beauty.

The nature-culture narrative of Boston Harbor and its islands also found firm footing in the city's and state's political circles. Boston mayor Kevin White, who held the office from 1968 to 1984, although mostly interested in the development potential of the inner-harbor waterfront, seldom missed an opportunity to refer to Boston's "historic harbor" or the city's "historic link with the sea."[16] Governor Sargent warned that Boston and, by extension, Massachusetts had for too long "turned its back on one of this State's . . . most important natural resources," the islands of Boston Harbor; the islands, he pointed out, were at the same time "richly woven into our heritage."[17] The change of rhetoric that went hand in hand with the islands' new role as cultural heritage and natural resource was solidified by the 1980s. By that time, a battle raged around the harbor about where the metropolitan area's new sewage-treatment plant should be built. Francis McCauley, the mayor of Quincy, used very clear words when speaking about the possibility that the plant might be sited on Long Island: if the plant was built there, the island would be "sacrificed"; if not, it would be "spared."[18] William Golden, Quincy's city solicitor who was instrumental in initiating the Boston Harbor cleanup and who was by then state senator for the district of Weymouth, agreed that siting the sewage plant on Long Island would be a "tragic waste of another harbor island."[19] A 1983 opinion piece in the *Boston Globe* used even stronger words: four of the harbor islands, it claimed—Deer, Moon, Nut, and Spectacle Islands, all former or current sites of waste-processing plants—"have been ruined for our times."[20] What fifty years before seemed like rational use of the islands was now deemed a waste and a ruin of the islands' potential.

Within only a decade, the islands went from Boston's dumping ground and its last vacant lot to places that it seemed almost a crime not to protect and cherish as natural enclaves and cultural heritage. The initially separate discourses about the harbor islands coined by the environmental movement and its proponents from grassroots to state level and by Edward Rowe Snow and other harbor boosters converged into one narrative of the Boston Harbor Islands as a storied wilderness.[21] The islands seemingly offered both pristine, undisturbed nature and a connection to the region's past defined by exciting maritime lore. When judged by their historical and scientific accuracy, both parts of the narrative proved equally problematic. The notion that the heavily disturbed nature of the harbor islands—shaped by human-induced erosion and the extractive economy of colonial Boston; their flora made unrecognizable by centuries of pasturage, logging, and planting with exotics; their salt marshes largely gone; and the shellfish communities of their tidelands depleted by pollution and intensive harvesting—was in any way pristine was as implausible as Edward Rowe Snow's version of the islands' history, presented as a series of sentimental and adventurous anecdotes that largely eclipsed the less romantic parts of the islands' actual histories and their close connection to the communities on the harbor shores. The resulting image of the harbor islands as an antidote to the ills of urban life, however, proved immensely attractive. Catering to the urban population's craving for a break from the city's noise and struggle, the islands offered an escape into a magical landscape that comprised natural beauty and an entertaining version of history. Since its invention this vision became instantly popular with the residents, the media, and the political representation of Greater Boston. The former gravel quarries, quarantine hospital grounds, and potential sites of residential communities entered the final third of the twentieth century as a historic wilderness.

Creating a Storied Wilderness

The perception of the Boston Harbor Islands as wild nature was based more on appearances than on the actual state of the islands' ecological communities. In 1969, as the plans for Expo 76 and for a city of the future in the harbor were still being discussed, the Special Commission on the Boston Harbor Islands came to the conclusion that the interests of Bostonians were best served by the public acquisition of the harbor islands under a program of conservation and recreation.[22] Part of the commission's proposal was the creation of a Boston Harbor Outer Sanctuary, comprising the four Brewster Islands, the two Calf Islands, and Green Island, and devoted almost exclusively to nature conservation. While the proposal was entirely in keeping with the traditional perception of the outer islands as wild outposts of untamed nature, from an ecological point of view it was hard to see what set them apart from the rest of the harbor islands. Over the years, they had been exposed to the same kinds of resource extraction, appropriation for public facilities, and human-induced changes in their flora and fauna as the harbor islands closer to shore. Rock from Outer Brewster had been used in the Boston area for building purposes; defense fortifications were built on Great and Outer Brewster Islands during the Second World War; all the islands had been deforested, along with their inner harbor cousins, by the European settlers in the seventeenth and eighteenth centuries and had been replanted—or had spontaneously overgrown—with exotics; when Green Island was overrun by rats in the 1960s, raccoons were introduced to keep the rodent population down.[23] Despite their rugged and undisturbed appearance, there was not much pristine nature left on the outer islands, and if any original New England wilderness was to be preserved there, it would have had to be re-created first.

The new way of perceiving the harbor islands as a romantic wilderness produced a change in how they were managed, preserved, and developed. Since the early 1970s, the only appropriate use for the harbor islands seemed to be conservation and recreation. The harbor-islands bill proposed by the Special Commission in 1969 narrowly missed approval by the Massachusetts General Court. Only after all mentions of possible housing and industrial uses of the islands were purged from the plan did the now exclusively recreation-oriented bill pass in 1970 in a legislature that was "more aware of the political sex appeal of environmental protection" than its predecessors.[24] The bill officially established the Boston Harbor Islands State Park, allowing the state to purchase sixteen of the harbor islands for recreational purposes through a $3.5 million bond issue.[25] In 1970, however, the park was more an abstract idea than a reality. The only islands the state owned at that point were Georges, Lovells, and Deer Islands, with Deer taken up by a prison and a sewage-treatment plant and therefore unavailable for recreation. The only island that had regular public boat access from the mainland was Georges. Simultaneously, Senator Ted Kennedy pursued the creation of a Boston Harbor National Recreation Area, a plan he first announced at the 1969 harbor-islands conference. The main stated goal of this plan was to provide recreational opportunities to all social classes of the Boston population: while criticizing the deteriorated state of the harbor environment, Kennedy stressed that he looked "for the day when all of the citizens of Greater Boston, both the underprivileged and the undeprived, can look forward to a weekend of camping and swimming on any of our islands."[26] The "Kennedy Bill" kept floating around for years: in 1976 a newly formed Boston Harbor Islands Advisory Commission was still debating its by then largely dormant legislative status and its primary focus on recreation instead of protection of natural resources. In 1979 Kennedy and Joe Moakley, by that time representing Massachusetts in the House of Representatives, made an attempt to push the bill through Congress, but their bid was unsuccessful.[27] Although the plan ultimately bore fruit almost twenty years later, it helped solidify the image of the harbor islands as a recreational area in the public's mind.

Meanwhile, the existing state park was slowly growing. In addition to Georges and Lovells Islands that formed the original park core, the state acquired Gallops and Green Islands, Outer, Middle, and Great Brewster, Calf and Little Calf, as well as the tiny Hangman Island in Quincy Bay in 1973 with the stated goal of using them for passive recreation.[28] In contrast to earlier plans for covering the islands with luxury hotels, glass domes, and marinas, the plans and practices of the 1970s were guided largely by an understanding of the harbor and its islands as places valuable because of their wilderness quality and because of the contrast they offered to the urban environment. The Harbor Islands Week, organized annually in the summer season by the state park, accentuated activities that designated the harbor as a romantic and extraordinary space replete with natural beauty and cultural heritage: the program included a turnaround of the tall ships and a Civil War muster at Fort Warren, but also tours of Lovells, Gallops, Grape, Bumpkin, and Great Brewster Islands that offered an undisturbed-looking natural scenery.[29] The whole event framed the harbor as a space where nature met history in a symbiotic way and the primary purpose of which was enjoyment and recreation.

The Boston Harbor Islands Comprehensive Plan, which was meant to guide the first years of the harbor park development, attempted to accentuate both the historical romance and the wilderness element of the islands' appeal. The document, prepared in 1972 by the Metropolitan Area Planning Council for the

Department of Natural Resources, formulated a development plan for the islands that remained in place until 1986. It was the plan's declared purpose to "recognize the unique character of each island by emphasizing its historical and natural attributes." The MAPC envisaged a combination of restored historical buildings and re-created wildlife areas on the islands, both to be enjoyed by visitors. The salt marsh at Thompson Island was supposed to be developed into a "major wildlife sanctuary" that, however, would be made accessible via a system of trails, floating boardwalks, and viewing platforms. The west head of Peddocks was to host a managed wildlife area for the black-crowned night heron, while the Brewsters were to become an enclave focusing almost entirely on marine environment protection. To contribute to the natural image of the islands, the plan recommended using materials "indigenous to the islands" such as stone or wood for the design of the facilities. As for history, not all of it had the same value to the park planners: while fort ruins on islands were to be restored and used for park purposes, other reminders of the long history that the islands shared with Boston were less welcome. The Long Island plan foresaw the restoration of the "visually and historically interesting" Fort Strong; the Long Island hospital, on the other hand, was in the eyes of the MAPC planners nothing more than an example of the "institutional misuse of the island."[30]

Despite its proclaimed devotion to preserving the islands' natural characteristics and to minimizing "potential adverse environmental impact of the development of recreational facilities," the 1972 plan was still to a certain extent affected by the grandiose development schemes of the late 1960s. The plan included building large visitor centers at Long, Peddocks, and Georges Islands, with the complex on Long Island set to include a swimming pool, tennis courts, a restaurant, an open-air café, and a dance hall.[31] However, missing funds and the complicated tangle of jurisdictions over the islands caused these development plans to remain on paper only. During the 1970s and the early 1980s, the islands' development was limited to clearing of trails, building of five new landing piers, and the opening of several campsites. As Kathy Abbott, by this time working as an environmental planner for the DEM, remarked, this innocuous way of developing the islands "seems to suit the 170,000 people who visit the islands each year just fine."[32]

The park grew out of the idea that the harbor islands possessed qualities unattainable in the city: wild nature, open space, calm, and tranquillity. This was what made them special and what made them popular, yet their very popularity threatened to destroy these qualities and the islands' uniqueness. By the 1980s, many of the park managers were clearly worried that the islands were in danger of being proverbially "loved to death." The islands were caught between two contrary forces: their appeal rested on their image as a romantic wilderness and on their insular quality, yet in order for these qualities to be enjoyed by the population of Greater Boston, they needed to be made as accessible as possible. The park management was thus walking the fine line of fulfilling the park's role as a metropolitan recreational area, but at the same time preserving the sense of the islands' uniqueness and isolation. Attempting to square this particular circle, the 1986 Boston Harbor Islands State Park Master Plan introduced a system of intensive-, moderate-, and low-level-use islands as well as numbers for individual island and park "saturation": the number of people who could visit the islands daily and annually without the islands losing their wilderness quality altogether. As a 1984 draft of the master plan noted: "Surveys of harbor island visitors indicate that while some improvements are desired . . . the fragile and subtle qualities of the islands are very much appreciated as they are. An unhappy scenario for the islands would see 'improvements' and growing numbers of visitors detracting from this environment and spoiling the islands for both human visitors and wildlife communities."[33] While attempting to make at least some of the islands more accessible for a larger number of people, the park managers also strove to make the islands appear more like a wilderness and less like a Boston suburb. Natural resources were to be not only protected but also restored: one of the 1986 plan goals was to support plants and wildlife on the islands "as they might have existed there before the 17th century."[34]

By the early 1980s, the harbor-islands state park was a relatively popular feature of the metropolitan system of open spaces: in 1980 some 180,000 people visited the park, with 60 percent of the visitors coming from Boston, 30 percent from elsewhere in Massachusetts, and 10 percent being out-of-state and international tourists.[35] The park seemed to be fulfilling its primary role as a leisure space for Boston, and park planners were looking for ways of boosting its popularity even more: after the implementation of the new 1986 master plan, they expected the annual visitor numbers to soar to a half million.[36] Before the harbor-islands park could become a true recreational hub, however, Boston and the state needed to address the elephant in the room: the harbor pollution.

The Boston Harbor Cleanup

The pollution of Boston Harbor caused by untreated sewage flowing into it had been an issue already in the first decades of the twentieth century.[37] At that time, it was regarded as mostly an aesthetic problem and not a serious health concern: a 1930 report of a special legislative commission established to investigate the discharge of sewage into the harbor found that the "harbor waters are in satisfactory sanitary condition."[38] A successive report of the same commission from 1935 already found the pollution problem steadily growing: by then harvesting shellfish was prohibited by the Health Department in most of Boston Harbor, unless the catch had been treated in a purification plant. The report identified the large number of combined sewer outlets—pipes that conducted sewage as well as storm water and that during rainfall discharged untreated sewage directly into the city's shoreline—as the main cause of the "objectionable conditions" in the harbor. Although the report concluded that the "condition of the Harbor is not such as to result in a nuisance," it nevertheless recommended the construction of sewage-treatment plants in the future.[39] In 1952 a treatment plant was built at Nut

Island, the outfall point of the southern branch of the metropolitan sewerage system; in 1968 the treatment plant for the northern branch at Deer Island followed.[40] Both facilities provided only primary treatment of the sewage, a method that allowed the heavier wastewater components to settle at the bottom of the tanks, while grease and oils rose to the surface and could be skimmed off. The rest would be then released into the harbor waters.

The treatment plants, however, did not make the pollution situation in the harbor much better. In 1970 Deer and Nut Islands were discharging 440 million gallons of partly treated sewage into the harbor every day, resulting in the closing of some of the beaches and in restrictions on harvesting shellfish. Besides, they did nothing to curb the flow of raw sewage from dozens of combined sewer outlets lining the harbor shore.[41] In 1969 the Boston Harbor Islands conference already treated harbor pollution as a serious problem that needed to be taken care of before any large-scale recreational development could take place. From then on, both the problem and its visibility grew rapidly. Whereas during the 1969 conference Frank Bergen of the Metropolitan District Commission (MDC) could still claim that the discharge of sludge into President Roads did not affect marine life at the outfall, only ten years later such a statement would have likely caused a public outrage.[42] A 1975 report by a city pollution-control body stated grimly that "the harbor waters along the beaches are only marginally fit for bathing" and that the inner harbor is "so polluted as to be disgusting."[43] The culprits were not only raw sewage discharges into the harbor—a 1977 MAPC newsletter counted 192 of them along the shore of Boston Harbor—but mainly the two sewage-treatment plants.[44] Both Nut and Deer Islands were often overwhelmed by the amount of sewage they had to deal with and were sometimes reduced to releasing the overload into the harbor waters untreated. Both plants suffered from frequent equipment breakdowns: a series of such events over a six-month period in 1982 sent 3.7 billion gallons of raw sewage down the harbor. Both were underfunded and more or less ignored by the MDC management, leaving it to their engineering teams to make do as best they could.[45] By the early 1980s, the harbor made the headlines every month, and the news was rarely good. Metropolitan-area newspapers reported about fish populations ridden with parasites and tumors, about beaches polluted by sewage during rainstorms, about the 452 million gallons of sewage and one hundred tons of sludge the harbor received every day, about the heavy metals and waste oils contained in the sludge, and about the high levels of DDT and PCBs polluting the harbor water and being digested by marine animals.[46] A 1986 study commissioned by the EOEA found that the combined effluent discharge from Deer and Nut Islands was the second highest in the whole of the United States and that the harbor's level of pollution compared unfavorably even to such notoriously polluted estuaries as the Chesapeake Bay or Puget Sound.[47] Few doubted that Boston Harbor fully earned its dubious recognition as "America's Dirtiest Harbor."[48]

It ultimately needed a series of lawsuits to bring the MDC out of its inertia. The founding myth of the cleanup took place at Wollaston Beach in Quincy, where the city's solicitor, William B. Golden, went for a run one morning in the fall of 1982. When he realized that what he at first held to be jellyfish was in fact filth washed on the beach—anything from spoiled fruit over used condoms to human feces—he was appalled. His initial disgust soon gave way to more constructive impulses, and in December 1982 he filed suit on behalf of the city of Quincy against the MDC in the Superior Court of Norfolk County. The year 1983 brought another lawsuit on the subject: the Conservation Law Foundation sued the MDC for water-pollution violations in Boston Harbor and the EPA for failing to perform its duties under the Clean Water Act.[49] The matter remained in the courts, under intense public and media scrutiny, for two more years: the number of lawsuits was augmented in January 1985, when the EPA sued the Commonwealth of Massachusetts and the MDC in federal court.[50] In 1985 U.S. District Court judge A. David Mazzone found the MDC liable for polluting Boston Harbor with millions of gallons of sewage and sludge.[51] The legal battle, however protracted, proved instrumental in fixing the pollution problem: the state broke up the unwieldy MDC in 1984 and created a new agency, the Metropolitan Water Resource Authority (MWRA), to manage Greater Boston's water supply and sewage system. In 1988 the harbor cleanup was set to begin: a mammoth project facilitated by state, federal, and city agencies and pushed forward by blue-ribbon panels like the Boston Harbor Cleanup Committee, established in 1983 by Governor Michael Dukakis and headed by the former governor Francis Sargent, and by citizen groups like Save the Harbor/Save the Bay, started in 1986 by William Golden, Paul Garrity, the judge in charge of the Quincy case, and the *Globe* columnist Ian Menzies.[52] The duration of the project was estimated at eleven years, its cost a staggering $6.1 billion.[53] In the course of the next ten years, the Nut Island plant was closed permanently, and the sewage from the southern branch of the sewerage system was transported through a tunnel to Deer Island. The Deer Island plant was rebuilt entirely, its capacity expanded and secondary treatment added. By 2000 the coliform bacteria counts in the previously affected parts of the harbor fell significantly, and the amount of sludge in the water fell to 20 percent of the 1988 levels. Even the project's cost proved lower than expected: $3.9 billion instead of $6.1 billion, due to low contract bids and cost cuts.[54] Boston Harbor was once again swimmable, both for fish and for human beings. Through the successful cleanup, the harbor gained a new meaning in the minds of Bostonians: it now became a place of environmental renewal, a space where marine ecosystems, after having been severely damaged by pollution, could recover and that transformed from a "harbor of shame" into an environmental success story.[55]

Several of the Boston Harbor Islands have played important roles in the history of the harbor's pollution and cleanup. Moon Island was the site of the city's first attempts to control the sewage flowing into the harbor in the late nineteenth century. Nut Island became a symbol of all that was wrong with the system of sewage treatment and of the harbor's appalling pollution. Deer Island, while having started out as one of the main pollution sources, eventually transformed into a figurehead of the success of the harbor cleanup, its huge digestion tanks becoming a new landmark of

Boston Harbor and the plant itself hailed as an engineering wonder. Long and Peddocks Islands were at various points considered as potential sites for the new sewage-treatment plant.[56] Even those islands that had never been contemplated as parts of the sewage disposal system were influenced by the cleanup: the clearing of the harbor waters made most of them into even more popular recreational spots than before. There was, however, no other place in the harbor where the old and new ways of using the harbor and its islands would clash so directly as on Spectacle Island. The surge of recreational uses of the islands, the persistence of perceiving the harbor islands as dumping grounds, the up-and-coming imagination of the harbor as a place of environmental renewal, and the reclamation of the islands "ruined" by former uses all came together poignantly at Spectacle's garbage-strewn shores.

The Sludge Wars

Spectacle's history did not lend itself easily to processing into tales of mystery and adventure. There was little romance in dead horses and piles of garbage, and even the harbor's master storyteller Edward Rowe Snow was able to find only thirty marooned woodsmen from the colonial times and one unexplained death that happened near the island when searching for exciting events in the island's past.[57] Similarly, Spectacle's flora and fauna seemed too disturbed and unoriginal and its garbage-dominated landscape too unappealing for it to pass as wilderness. Most harbor-island writers and park planners thought that Spectacle could be turned into a "useful" island once again and that its usefulness would be defined by providing opportunities for recreation and an escape from the city—but it would be in need of a complete makeover. Alan Mikal, the author of a 1973 Boston Harbor guidebook, although one of the few people who saw natural beauty in Spectacle as it was in the early 1970s, nevertheless asserted that the occasional visitor would be able to savor Spectacle's charm only if he or she "disregards the island's unsavory past."[58] The MAPC planners responsible for the 1972 Boston Harbor Islands Comprehensive Plan took the same basic view of the island's history, with a little technology enthusiasm thrown into the mix. They suggested that Spectacle had "excellent potential" for recreation and that, once "reclaimed," it could host a five-hundred-boat marina, picnic areas, playfields, and a swimming beach. The reclamation of the landfill would even have been temporarily integrated into the island design, with boardwalks constructed across the dump to allow observation and with interpretive signs explaining the reclamation process. Ultimately, though, the island would emerge as a green recreational paradise and with the addition of a final layer of topsoil, "the signs of the island's 100 years of abuse would become only memories."[59] Obliterating Spectacle's past seemed the only way for the island to have a future.

The 1986 Boston Harbor Islands State Park Master Plan that followed the 1972 plan as the comprehensive document outlining the islands' development and management was even more explicit in its endeavor to cover Spectacle's "unsavory" history. While the 1984 draft version of the plan contained direct references to the island's specific natural conditions and included a map that showed in detail the location of the "eroding refuse coastal banks" or a "beach composed of sand, stone, glass, metal, pottery and brick," in the 1986 final master plan all mentions of refuse and erosion were left out. References to the refuse showing in the island's features were replaced by the more opaque term *landfill area* and the island's contours rendered an unobtrusive, soft shade of green. The resulting version clearly tried to downplay and obscure the island's environmental problems in order to present it as an opportunity and a wonderful resource rather than a nuisance and a leaking former garbage dump.[60] Spectacle Island seemed to be on a straightforward track to become, sooner or later, one of Boston Harbor's recreational hubs—if it had not been for sludge.

Of all the difficult steps on the way to a clean Boston Harbor, sludge treatment was one of the most bothersome. In 1982 the Nut and Deer Islands sewage-treatment plants produced altogether around 174 million gallons of sludge, the solid matter that settled at the bottom of the sewage tanks during primary treatment.

Pumped to anaerobic digesters for stabilization, it was then released into the harbor.[61] A major pollutant containing high concentrations of toxins, pathogens, and heavy metals, the release of sludge into the harbor waters caused significant ecological damage and prevented the harbor from becoming much else than the city's waste receptacle.[62] One of the more important decisions the newly established MWRA was confronted with was to select a method of sludge disposal—composting, incineration, or processing into fertilizer—and to choose a site where the sludge treatment would happen. The latter one, however, proved trickier than the sludge treatment itself. Sludge-processing plants were perceived by the public as smelly nuisances, and when several communities in the Boston metropolitan area found out that they were being considered as future sites of the sludge plant, passions ran high.[63] Quincy, Stoughton, Revere, Walpole, and other towns that felt threatened by the sludge plant—although reminded by the MWRA board member Robert Ciolek that they all sent their sewage to Boston Harbor and that if they were part of the problem, they should be part of the solution—were suddenly united by a common cause.[64] They had a common problem and soon found a common solution. Among the proposed sludge-processing sites, they discovered one they all could agree on: Spectacle Island.

The sludge battle pitted two camps against each other: on the one side, there was the state park management, the conservationists, and harbor boosters who saw Spectacle Island as a place predisposed alone by its quality of being an island for recreational uses and for becoming a jewel in the crown of the Boston Harbor Islands State Park. On the other side, there were the residents and political representatives of the towns under consideration for the sludge-processing plant: in their eyes, Spectacle was a place spoiled beyond redemption, a place predestined by its history of waste disposal to accommodate yet another such use. Theirs was a not-in-my-backyard mentality coupled with a path-dependent way of thinking that saw the harbor and its islands in their traditional role as the metropolitan region's dumping ground.

When they spoke about Spectacle, it appeared as if the two opposing groups were talking about two completely different places. The antisludge coalition, united since the summer of 1988 under the surprising acronym OASIS (Organization Advocating a Spectacle Island Site), painted Spectacle in the darkest colors. Steven Fradkin, board member of both OASIS and Stoughton against Sludge, called the island simply "a trash heap overgrown with grass."[65] When OASIS organized two cruises to Spectacle in late August 1988, its members made sure to supply the reporters present with a hearty portion of memorable quotes. State representative Marjorie Clapprood of Sharon climbed Spectacle's trash cliff to plant a flag on top of it and sarcastically declared the island "the perfect site for the sludge." Quincy city councilor and fellow OASIS member Michael Cheney declared that the island "is simply too polluted to ever be used as a park." As for its current state, OASIS members claimed that the island had been identified as one of the worst sources of pollution in the harbor, and, as Cheney observed, there was no wildlife in sight because "the island is even too disgusting for rats."[66] Cheney also wrote to the secretary of environmental affairs and MWRA chairman James Hoyte, arguing that Spectacle, "unlike most other harbor islands, has no significant history" and can therefore, given its past uses, continue its career as a dumping ground.[67]

Unsurprisingly, their impressions were not corroborated by those who saw something else in Spectacle than a "rat-infested, disease-ridden wasteland."[68] Planners of the Boston Harbor Islands State Park were eyeing Spectacle as a potential "intensive-use island" that could be transformed into an "attractive and enjoyable island park."[69] The 1986 master plan stated that although trash could be found on most of the island and although its cliffs were eroding, there were no toxic wastes to be found on the island.[70] A thorough environmental investigation of Spectacle conducted in 1988 by the MWRA revealed a highly disturbed environment dominated by the presence of the landfill, yet teeming with life. The island was overgrown with thick vegetation in which shrubs were slowly replacing grassy species in the usual way of plant succession. The island was home to a small salt marsh where vital wetlands vegetation thrived, including several species of cordgrasses and reed grass of the *Phragmites* genus. In its tidelands, the common North Atlantic species of mussels were abundant. During their site visit, ornithologists from New England Environmental observed twenty-nine species of birds on the island: among others, the island appeared to be a nesting and breeding site for herons.[71] Finally, the New England chapter of the Sierra Club maintained that Spectacle, even in its present state, was used as a recreational area by the surrounding communities and that it was in fact valuable as it was, even without a state park–planned makeover: "It is not a smelly, burning, horrible dump," the Sierra Club wrote in their position paper sent to the EOEA. "It is pleasantly verdant and offers a cliff of archaeological interest."[72]

Yet others, even though not directly participating in this debate, voted with their feet. Although not part of the park system and

FIGURE 25. Spectacle Island in the late 1980s. While some dismissed it as "even too disgusting for rats," others praised its potential for recreation. Even as an uncapped landfill, the island attracted relatively large numbers of visitors.—*Photo by the Central Artery/Third Harbor Tunnel project. Courtesy of MassDOT.*

officially closed to visitors, Spectacle was apparently used illicitly as a recreational spot throughout the 1980s. Its location near the main shipping channel made it accessible, and its lack of policing invited activities not allowed on islands overseen by park staff, such as alcohol consumption and shooting guns. Spectacle's uncontrolled, unregulated wilderness and the occasional treasure yielded by the garbage dump may have contributed to its appeal. How many people used Spectacle in this way was unknown, but some of the harbor volunteers suspected it was more than visited some of the islands within the park system.[73]

Ironically, it was the not-in-my-backyard principle that ultimately prevented the construction of the sludge plant at Spectacle. In August 1988, at the groundbreaking ceremony for the new sewage-treatment plant at Deer Island, the Massachusetts Senate president, William Bulger, declaring to everyone's amusement that he spoke "as an ordinary, humble citizen," announced his opposition to the Spectacle sludge plan. He equaled the building of the plant to "sacrificing" the island and called Spectacle a "precious resource" that should be kept open for recreation.[74] The humble citizen's words resonated in the local press all around Boston Harbor and made many an OASIS member stop in their tracks.[75] One of Walpole's selectmen even wondered publicly whether Walpole's opposition to the plant within its town limits was wise, as it might upset the powerful senate president.[76] Despite appearances, Bulger's and OASIS's motivation fed on the same sentiment: William Bulger represented a district that included South Boston where his own home also stood, less than two miles from Spectacle, which once again threatened to become a malodorous nuisance. His reasons may not have been those that he stated, but Bulger's opposition, combined with the MWRA's doubts whether the former landfill provided a stable-enough ground for the plant's construction, the DEM's plans for Spectacle to become parkland, and the lobbying of conservationist groups such as the Sierra Club, effectively stopped the consideration of Spectacle as a site for the sludge plant.[77] The plant was eventually built in Quincy, turning sludge into fertilizer pellets and sending them to citrus groves in Florida and golf courses across New England.[78]

It was not NIMBY rhetoric, though, that hindered turning Spectacle into Greater Boston's dumping ground yet again. Senator Bulger, when opposing the scheme, chose to argue his case by means of the new harbor-islands narrative that had already become the standard way of looking at the islands. Although his motivation probably lay in the self-serving not-in-my-neighborhood logic, he nevertheless praised Spectacle in the language that became universally accepted when speaking about the harbor islands: the language of preservation and the intrinsic value of the islands' nature and open space. Even those who, like the Quincy councilor Michael Cheney, had reasons to see the harbor islands as dumping grounds first felt the need to establish that there was no nature and no history—at least not the right kind of original New England nature and no impressive heritage of forts and shipwrecks—on Spectacle before it could be used as a site for processing Boston's trash.

The rule of recreation over Boston Harbor was not absolute, though, even in the late twentieth century. Although recreational uses enjoyed great prestige and were at the top of the imaginary list of desirable uses of the harbor, they shared the harbor space with a wide variety of other activities. The port of Boston, although reduced in size, still shaped the harbor by claiming several waterfront portions in South and East Boston and by dredging the harbor's silting channels to a depth required by the behemoth container ships. The slowly clearing harbor waters encouraged harvesting shellfish, as evidenced by complaints about the great number of lobster pots that obstructed navigation.[79] Above all, the harbor never ceased to play, to some extent, the role of Boston's ultimate sink. When the MDC and later the MWRA were looking for a place to put the new sewage-treatment plant, almost all of the sites under consideration were harbor islands.[80] What made the islands favored locations of wastewater treatment was not only their convenient position in the harbor into which the effluent from the plant would be released but also their historical roles and the persistent imagination that saw them as isolated places fit for uses that would be bothersome on the mainland. Only an appeal to natural or cultural merits could prevent an island from being converted into an unpopular municipal facility. When Boston mayor Raymond Flynn proposed to build a prison on Peddocks Island in 1990, the EOEA secretary, John DeVillars, argued against it on the grounds that Peddocks possessed "many unique natural and cultural resources," including a wildlife refuge, a six-thousand-year-old Native American archaeological site, and an Endicott-period fortress eligible for the National Register of Historic Places. He suggested Long Island instead, which, although equally scenic and equally crowned with an Endicott-era fort, also featured the sprawling complex of the hospital for the chronically ill. This part of the island was suggested as an appropriate location for the prison, since the hospital was not deemed heritage worth preserving.[81] The site selection for the sewage-treatment plant followed similar rules: the plant eventually went to Deer Island, mainly because of its low projected cost compared to other alternatives, but also because of the general perception of the island as having been already ruined for recreational purposes by the old sewage plant and the Suffolk County prison.[82]

Even the complex and costly Boston Harbor cleanup did not put an end to treating the ocean as a sewer: in a sense, it merely meant a longer pipe. While previously the outfall pipe from Deer Island discharged directly into the President Roads channel, making a substantial portion of the effluent wash back into Boston Harbor with the tide, the newly constructed outfall pipe brought the treated sewage nine and a half miles out into Massachusetts Bay.[83] The idea governing wastewater disposal was still externalization—and the notion that if Boston Harbor was not as vast and infinitely able to absorb trash as it seemed a hundred years ago, the Atlantic Ocean might be. In fact, the same principle had been by then used for almost fifty years elsewhere in Massachusetts Bay. Twenty-two nautical miles east of Boston lay Massachusetts's "open-water disposal site," a circular area of the ocean floor two miles in diameter where the material dredged from ship channels and marinas along the Massachusetts coast was disposed of. Known as the Foul Area, the site had been in use since the 1940s and until 1977 accepted low-level radioactive waste, toxic industrial waste, and explosives; after 1977, it was used only for dredged material. Located next to the eastern slope of Stellwagen Bank, Massachusetts's only National Marine Sanctuary and an important feeding ground for the critically endangered North Atlantic right whale, the presence of the Foul Area raised concerns of marine biologists about possible dispersal of pollutants.[84] Despite its location next to a marine reservation, the site continued to operate. Boston did not abandon the use of its coastal waters as a sink; it only moved the outfalls and the landfills farther away.

The Boston Harbor Islands National Recreation Area

When Congress declared the Boston Harbor Islands a national recreation area in November 1996, it was the culmination of a decades-long effort that reached from the grassroots to the federal level. The names most often associated with the new island park were those of Senator Ted Kennedy and Representative Gerry Studds. Both longtime proponents of environmental protection and of creating a national park in Boston Harbor, they once again took up the dormant park cause in Congress in the early 1990s. In 1992 Studds sponsored federal legislation providing for a special study meant to determine the suitability of the Boston Harbor Islands to become a national park. Legislative decision making, however, was just the tip of the iceberg. The feasibility study, completed in 1994, was the result of fourteen months of the National Park Service's consultations with hundreds of state and city agencies, nonprofits, business organizations, island owners, and other stakeholders in Boston Harbor. When the Boston Harbor Islands were found a worthy and feasible addition to the national park system in 1996, it was above all a success of a campaign led by the harbor nonprofit organizations. Groups like the Boston Harbor Association, the Friends of the Boston Harbor Islands, or Save the Harbor/Save the Bay helped create widespread support for the park and influenced the way the park and its management plan looked in the end. The creation of the park also confirmed the new role that Boston's harbor islands now filled for their city.[85]

The 1994 feasibility study subscribed fully to the now well-established narrative of the islands as naturally and culturally significant places. Informed by environmental consciousness, it did not eclipse even the islands' past "unsavory" uses and understood them as potential educational resources that could teach visitors about the human impact on the environment and about historical attempts to deal with issues associated with urban growth. Nonetheless, the pieces of the islands' history that the park planners related to most were the late-nineteenth-century attempts of Frederick Law Olmsted and Charles Eliot to include the harbor islands in Boston's metropolitan park system. Choosing two of America's most celebrated landscape architects as patron saints of the new park was understandable: it made the present park planners seem like heirs to their vision. It also suggested the context in which they wanted to see the new park: not just as a separate natural area but as an integral part of a complex larger whole, the long-lost and finally regained piece in the jigsaw puzzle of Boston's metropolitan open spaces.[86]

The study also picked up a theme running through much of Boston's history: that of linking Boston's fortunes and the city's economic and social well-being to the physical state of its harbor. Nineteenth-century merchants and sea captains wondered if Boston's decline as a port had to do with the shoaling of its harbor. Many 1960s Bostonians were convinced that the best chance their city had at reinventing itself and becoming attractive for business and affluent residents was to expand into the harbor and onto the islands. Now, city and state agencies as well as business, civic, and advocacy communities argued that it was only through reconnecting with its historic harbor that Boston would claim its rightful place as a world-class city. It was only the nature of the connection that had changed: now it meant ensuring pedestrian access to the redeveloped waterfront, making the harbor islands easy to reach by water transportation, and integrating the harbor and the islands into the residents' lives as a vast and friendly urban green space.

"Of the many past uses the harbor islands embrace, recreation is arguably one that offers realistic prospects for the future" was the feasibility study's remarkably sober summary of the continuing utilitarian nature of Boston's relationship with its harbor and islands. Although, on the one hand, acknowledging the islands as places unique and valuable in their own right, the many actors involved in drafting the feasibility study proposed uses of the islands that were as ever governed by the city's current needs. The law establishing the Boston Harbor Islands National Recreation Area pledged to conserve, protect, and maintain "the scenic, historical, cultural, natural, and scientific values of the islands" not for their own sake but for them to be enjoyed by the public. The steeply growing population of the metropolitan region required more and more opportunities for leisure, some of which the harbor islands could help meet. The public enjoyment of the harbor park also had an important social component: the islands were to be made accessible to people of all classes and backgrounds, especially those who could not afford exotic vacations and were dependent on recreational opportunities close to home. Thus, decades of rising environmental awareness, efforts of local advocacy groups, population growth of Greater Boston, increasing demand for leisure, and the exceptional success of a narrative interpreting them as scenic and historic places helped turn the Boston Harbor Islands into Boston's largest urban park. To live up to their new role, however, many of the islands needed to be altered and "improved" accordingly. Nowhere in the harbor was this makeover more striking than on Spectacle Island.[87]

Building a New Spectacle

When Spectacle's long-awaited rebuilding into a park finally came, it was a triumph of the new imagination of the harbor islands as places of leisure and beauty. It was, however, also the island's latest appearance in its centuries-old role as the universal solution to Boston's problems, one final incarnation of the "got a problem" mentality with which Boston had approached its harbor islands ever since the first smallpox victim was ferried over to Apple Island in 1702.[88] Spectacle's emergence as a "jewel" of the Boston Harbor Islands park in the first years of the twenty-first century was facilitated by one more instance of its use as Boston's dumping ground.[89]

The 1980s and the 1990s were a time when Boston, ever a city bound on large-scale reshaping of its environment, remade itself yet again. Apart from the gargantuan undertaking of the Boston Harbor cleanup, the Commonwealth of Massachusetts and the City of Boston embarked in the early 1980s on another multibillion-dollar megaproject: the rerouting of the Central Artery, the freeway that had since the 1950s cut Boston's downtown in half, into a tunnel and the construction of a third tunnel under the inner harbor. The Central Artery–Third Harbor Tunnel project, CA/T for short and colloquially known as the Big Dig, had been in the planning phase since the beginning of the 1980s. The project was approved by the Federal Highway Administration in 1985, and after six more years of planning the ground was finally broken in 1991, with completion envisioned for 1998.[90] In the years that followed, however, the phrase that became most commonly associated with the CA/T project was "plagued by delays and cost overruns." Instead of 1998, the Big Dig was completed in 2007, and its costs soared from the initially projected $2.6 billion to nearly $15 billion, with more than $3 million spent every day during the peak construction years between 1998 and 2001.[91] For more than fifteen years, the Big Dig turned downtown Boston into a vast construction site, building more than 160 lane miles of roads and pouring 3.8 million cubic yards of concrete into Boston's earth—enough to build a three-feet-wide sidewalk from Boston to San Francisco and back three times over.[92] Despite its shortcomings, the finished project provided better access to Logan Airport and changed the face of downtown Boston completely: it alleviated traffic congestion, eliminated the looming "Green Monster" structure of the elevated Central Artery and replaced it with the pedestrian-friendly Rose Kennedy Greenway, and reconnected the waterfront district with the rest of the city.

Among other overwhelming statistics, the Big Dig generated about 16 million cubic yards of excavated and dredged soil, clay, rock, and sediment.[93] Faced with this towering mountain of dirt that would have been enough to fill sixteen stadiums to the brim, the project management and the Massachusetts secretary of transportation Fred Salvucci soon began casting about for a place to dispose of it. Early studies in 1983 assumed that the material would be trucked to a number of inland sites, while by 1987 the Massachusetts Department of Public Works (DPW) was reviewing a number of quarries, landfills, tidal flats, and ocean dumping sites in and around the metropolitan area for the purpose.[94] Most of them, however, had any number of drawbacks: mainland disposal sites could not accept salty dredged material for fear of groundwater contamination. Governors Island Flats that were also being considered could have absorbed even more fill than the Big Dig generated; this, however, would have meant the destruction of protected marshes and mussel beds. More and more, Spectacle Island began to emerge as an ideal site for the spoils. Putting "Salvucci's dirt" on Spectacle, it seemed, would have made everyone happy. The Boston Transportation Department supported the island site: the excavate would be ferried to the island on barges, saving the city thousands upon thousands of trucks thundering through its streets, causing traffic congestion and upsetting residents.[95] Lorraine Downey, the director of the Boston Environment Department, voiced her satisfaction with the fact that capping the island would put an end to pollutants leaching from the former landfill into the harbor waters.[96] The DEM was equally pleased: the 1986 master plan that slated Spectacle for development into one of the major recreational centers of the state park stated at the same time that the island, especially its landfill part, would require intensive stabilization before park development could commence. The master plan even mentioned the possibility of acquiring CA/T excavate for this purpose.[97] Now that the CA/T project management came offering all by themselves the Big Dig dirt to cover Spectacle, the DEM greeted the proposal with enthusiasm.[98] Finally, the CA/T project managers were relieved to have found a solution to one of their most pressing problems: Mervin Mirsky, a top engineer with the construction consortium Bechtel/Parsons Brinckerhoff, the project's main contractor, was quoted as saying that "using Spectacle is practically a do-or-die proposition for us."[99] Martha Bailey, project manager with the DPW, put matters even more bluntly: "Someone asked me, 'What would you do if you can't get Spectacle Island or the foul area?' . . . My answer was: Suicide. There aren't really any alternatives."[100]

As soon as they decided on Spectacle as the primary disposal site, the agencies involved started portraying their plan as an environmental blessing and a sort of ecological "rags-to-riches" story. Governor Dukakis framed the DEM's intent to build up Spectacle into a recreational hub as a story of redeeming the environmental sins of the past. "Spectacle," he commented during his announcement of the new state park master plan in December 1984, "is an opportunity to take people's biggest mistake on the Harbor Islands and turn it into a masterpiece."[101] While past uses had ruined the valuable resource that Spectacle was, along with the other harbor islands, the more enlightened present now had the opportunity to "reclaim" and "improve" it. *Access*, the CA/T newsletter informing metropolitan-area residents about the progress of the project, ran an article about Spectacle in 1988 titled "Deadly Dump." In it the CA/T project presented the capping of Spectacle as an act of public good and an essential win-win situation for all parties involved: the CA/T project, the Boston public, and the environment.[102] An ecological disaster could now be reversed and a harbor island restored to its best use as a recreational spot.

Despite their rhetoric, the environment was not one of the CA/T managers' and state agencies' greatest concerns when it came to Spectacle. Overjoyed by finding a seemingly simple solution to a difficult problem, the state and the city overlooked a number of both state and federal environmental laws. In 1988 the Department of Public Works proposed Spectacle as the preferred disposal site for 12 million out of the expected 13.5 million cubic yards of excavated material: this would have enlarged the island by 136 acres.[103] The DPW's satisfaction, however, was nipped in the bud by the New England EPA and the Army Corps of Engineers. Both agencies pointed out that filling more than 130 acres of Boston Harbor violated the EPA's wetlands-protection guidelines and failed to comply with the Clean Water Act. The EPA's regional administrator, Julie Belaga, and the Corps of Engineers' colonel Daniel Wilson, during a meeting with DPW commissioner Jane Garvey in March 1990, voiced a suspicion that placing all the fill in the wetlands, mudflats, and waters around Spectacle was guided merely by convenience and not by concern for the aquatic ecosystem.[104] The plan also failed to include any landscaping or park construction on the island beyond the dumping of the fill, provoking a lawsuit from the Friends of the Boston Harbor Islands and Save the Harbor/Save the Bay advocacy groups.[105] In response, the future island design shrank as quickly as it had blown up before: first to 9.2 million cubic yards and 103 acres of harbor filled and only a few months later to 3 million cubic yards and 8 acres of fill.[106]

One of the sites around Boston that received the dirt originally meant for Spectacle was the Gardner Street landfill in Boston's neighborhood of West Roxbury. The landfill, which occupied a quiet bend of the Charles River, had between 1954 and 1980 received solid waste from all areas of Boston. During the operation of the South Bay incinerator from 1959 until 1975, the landfill was also the final station for ash from the incinerator as well as for materials that could not be burned there. Over the years, the mound of waste rose to 5.5 million cubic yards, covering some 90 acres of West Roxbury land.[107] Since 1995 the city of Boston was considering capping the inactive landfill with CA/T fill and constructing a

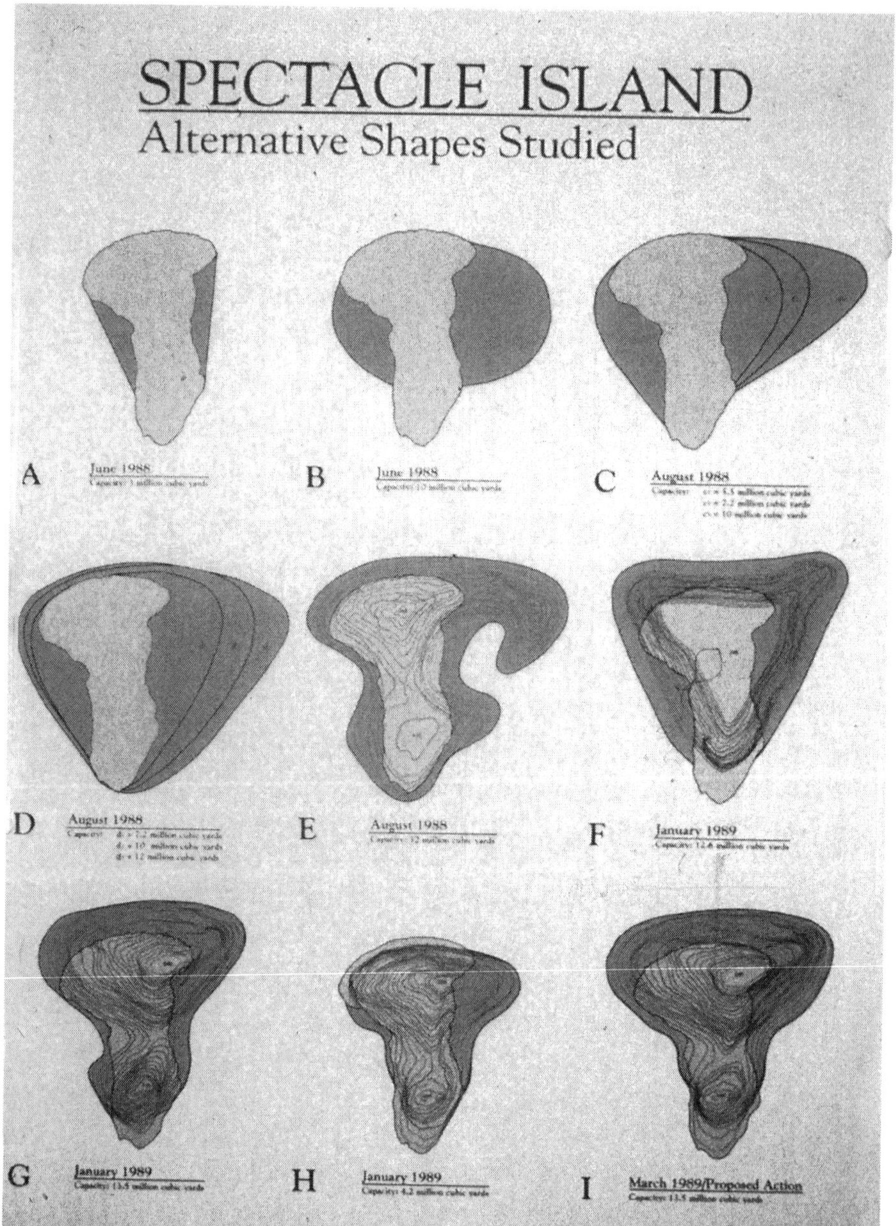

FIGURE 26. Alternative shapes proposed for a reconstructed Spectacle Island in 1988 and 1989. In their struggle to find a place for the excavate the Big Dig generated, its managers overlooked a number of environmental laws protecting wetlands.—*Image by the Central Artery/Third Harbor Tunnel project. Courtesy of MassDOT.*

park on it.[108] The size and the design of the proposed park looked like a blueprint for the development that was going on at the same time at Spectacle Island. At 98 acres, the park on top of the Gardner Street landfill was only 5 acres smaller than the fill-enlarged Spectacle, and its planned uses could have passed for a mainland copy of the island design. The old landfill was to be covered with layers of sand, clay, and topsoil; the city proposed building a park atop the trash heap, with sport fields, walking paths, a natural study area, newly planted trees, a picnic area, and lookouts with a view of the river at the foot of the reused trash hill.[109] The plan was lauded in the press as "turning a filthy environmental hazard into a beautiful city park," in language that made the impression of having been taken straight out of the Spectacle coverage.[110] In size, origin, and design, the two parks were almost identical; the only substantial difference was in the amount of attention they received. While the park construction at Spectacle Island was followed in dozens of articles in both Boston and national media, the makeover of the Gardner Street landfill into Millennium Park was covered only by the local *West Roxbury Transcript*. When Boston mayor Thomas Menino attended the Millennium Park groundbreaking ceremony in 1999, he praised the future park as a neighborhood asset. Six years earlier, Secretary of the Interior Bruce Babbitt toured the Boston Harbor Islands, including Spectacle, and pronounced them a "national treasure" fit to become part of the national park system.[111] Millennium Park was missing one key factor that made Spectacle prominent and interesting: it was not surrounded by saltwater. While Spectacle was one of the now-coveted harbor islands and therefore automatically regarded as possessing tremendous potential for natural beauty, scenic views, and recreation, Millennium Park was regarded as nothing more and nothing less than a reclaimed garbage dump, a quiet patch of green in the suburbs. The island location made all the difference: it elevated one of Boston's landfills to the position of "one of the city's more desirable recreation destinations," while an almost identical trash heap turned park without the benefit of the ocean vistas remained known only to West Roxbury locals.[112]

Before city officials even started contemplating turning the Gardner Street landfill into Millennium Park, the construction work at Spectacle Island was already going ahead at full speed. In late 1992, the first barges carrying material excavated from the future harbor tunnel arrived at Spectacle, and by 1993 the island was transformed into a busy construction site.[113] At first, the containing of the landfill and the depositing of the fill was progressing at a fast pace, with most of the material already placed on the island by 1995.[114] After that, however, years of interdepartmental feuds, cost cuts within the CA/T project and resulting design changes, and technological problems with the drainage system on the island kept pushing the project completion date further and further back. The erstwhile finishing year of 1997 soon became 2000, and each new expected completion date brought new delays instead.[115]

FIGURE 27. Spectacle Island during the construction phase in November 1993. In order to become a park, the island's original features were obscured along with the landfill. —*Photo by the Central Artery/Third Harbor Tunnel project. Courtesy of MassDOT.*

Although often described as "reclamation" or "rebirth," what happened on Spectacle could most accurately be described as building an island up from scratch.[116] The island had been stripped of vegetation, had sea walls built around it to prevent erosion, and before the capping could even commence had its two drumlins graded almost to the waterline.[117] The disorderly wilderness that had thrived on the original island had to go and make place for a more conventional, manicured, and controlled version of nature. Since Spectacle was to be reincarnated as a picture-perfect park, reminders of its past as a working landscape were largely unwelcome. There were hardly any of the factory, house, or wharf structures left, but the tall concrete chimney of the Coleman plant still stood in the saddle between the two island elevations. In 1983 the DEM's state-park newsletter—perhaps somewhat grudgingly—listed the seventy-foot smokestack as one of the harbor's landmarks, along with the Nixes Mate beacon and Boston Light.[118] After the construction of the new island began in 1993, however, the chimney was torn down, allowing the island to be reimagined as a visually pristine natural space.

FIGURE 28. Spectacle Island after capping and landscaping in 2002, its man-made environment set to become one of the park's intensive-use hubs.—*Photo by the Central Artery/Third Harbor Tunnel project. Courtesy of MassDOT.*

In 2006, after years of delays and public expenditures of $174 million, the first boat with visitors from Boston docked at Spectacle. The new island did not have much in common with the old one beside its name. Having absorbed 3.5 million cubic yards of Big Dig dirt, its profile had been raised by some 60 feet, so that its highest point on the northern drumlin was rising 157 feet above the sea level.[119] An elaborate drainage system had been installed in its core to capture pollutants leaking from the dump underneath the island's new look. Twenty-eight hundred trees and twenty-six thousand shrubs had been planted on its graded slopes, and the island had been seeded with native grasses.[120] A visitor center, dock, and sandy beach had been built on the island's west shore. Five miles of walking trails had been laid out to offer views of the harbor and of the Boston skyline.

Those who arrived on Spectacle Island on its opening day, the rainy Saturday of June 24, 2006, voiced their amazement at the island's restoration.[121] What they saw was undeniably an environmental success story: a source of harbor pollution capped and transformed into a green and pleasant park. What they also witnessed, though perhaps unwittingly, was a triumph of a certain kind of narrative about Spectacle and the Boston Harbor Islands: one that saw them as places best fit for recreation and enjoyment of natural beauty, places valuable in their own right, and places that were fundamentally different from the city.

EPILOGUE

On May 7, 2016, the Boston Harbor Islands National and State Park entered a new tourist season in a fine shape. Celebrating the twentieth anniversary of the national park, the Boston Harbor Islands Partnership, a coalition of twelve federal, state, and city agencies and nonprofit organizations jointly managing the park and including most prominently the National Park Service, the Massachusetts Department of Conservation and Recreation, and the Boston Planning and Development Agency, put together a program framing the harbor islands as a vast, vibrant urban playground. Activities were centered mostly on the park's three intensive-use islands, Spectacle, Georges, and Peddocks, with program points designed to fit the individual islands' resources and image. During the whole summer, Peddocks, designated in the park's *General Management Plan* as an island with "natural features emphasis," hosted activities accentuating the park's natural resources: kayak trips around the island, a class explaining the basics of saltwater fishing, and a ranger-led tour educating children about the "creatures that live below the tide line."[1] Spectacle, as a "managed landscape," offered a program that could be expected in an urban park: every week the island's visitors had the opportunity to join an open-air yoga class, fly kites, play giant Frisbee, and enjoy a sunset clambake. On Georges, designated as an area with "historic preservation emphasis," the activity concentrated on the island's cultural heritage: in addition to a Civil War Dress-Up, a Civil War Walk, and Civil War Lawn Games, visitors, especially the underage ones, could build a fort, storm the fort (every Saturday and Sunday at 11:15 a.m.), and naturally listen to the legend of the Lady in Black. Three mainland gateways at Boston's Long Wharf, in Hingham, and in Hull provided regular boat service to the three principal island parks, with less frequent connection to Thompson, Grape, Bumpkin, and Lovells Islands, as well as to Boston Light throughout the summer season.[2] The park's website and its newsletter, the *Harbor Islands Times*, gave the visitors an overview of the weekly activities, special programs, and boat-departure times, along with appealing images of happy children and adults engaging in park activities in idyllic island settings against the backdrop of Boston skyline in the distance.[3]

State and federal agencies such as the Department of Conservation and Recreation and the National Park Service were not the only organizations active in the harbor. Boston Harbor Now, the largest harbor nonprofit, worked together with the agencies, offering 426 free programs on the harbor islands and on the waterfront in 2017, in addition to organizing free boat tours and harbor events. Citizen advocacy groups like Save the Harbor/Save the Bay or the Friends of the Boston Harbor Islands also organized their own harbor cruises, trips to Boston Light, and tours of the islands' natural and cultural resources.[4] Although running on different boat schedules, all of them shared the same basic understanding of Boston Harbor and its islands as a consummate recreational area. This perception was first put in words by the establishment of the state park in 1970 and was reinforced in 1996 by the creation of the Boston Harbor Islands National Recreation Area. The Boston Harbor Islands of the twenty-first century became places defined almost exclusively by their natural beauty and historicity and by their recreational potential and uses. The new image and accessibility of the islands made it possible for thousands of day-trippers to relate to the islands through leisure activities, experiencing them as places of relaxation and recreation.

These new uses also created a whole new harbor class: people who engage with the harbor as often as the stevedores and fishermen of old, but in a different way. The working lives of park rangers, boat operators, nonprofit managers, and volunteers take place in and around the harbor, but are now governed by the demands of recreation. This new harbor class has developed its own understanding of the harbor, its own harbor narrative in which mystery and adventure mingle with ecological renewal and leisure. The Brewster Islands are no longer the "dreaded lee shore" they once were, but rather an "unspoiled, genuine wilderness" and a "jewel in the Harbor Islands crown," best used as a wildlife sanctuary and a playground for recreational kayakers and divers.[5] Georges Island is no longer one of the islands guarding the Narrows, an important shipping channel, and the site of one of the harbor defenses, but first and foremost the seat of the Lady in Black. Spectacle Island has left behind its past as a "trash heap overgrown with grass" in favor of a present as a beloved urban park and a symbol of harbor renewal.[6]

Edward Rowe Snow has rightfully claimed a central place in the new harbor mythology as the founding father of the recreational imagination of the harbor and, indirectly, of the state and national park. Authors of harbor islands guidebooks such as Christopher Klein have praised Snow as the father figure of the movement for preservation of cultural and historical resources of the islands.[7] In 2000 a memorial plaque for Edward Rowe Snow, paid for by citizen donations, was unveiled on Georges Island; the annual Edward Rowe Snow Day organized by the Friends of the Boston Harbor Islands has commemorated his birthday every August since 1980 with events on Georges Island. One of the boats that ferry passengers between Boston, Hingham, and Hull is named in his honor. All this revering is strangely appropriate: Edward Rowe Snow's activities had indeed been instrumental in turning the islands into what they are now: Boston's national park and urban open space.

The narrative that the National Park Service, the main custodian of the park, tells about the Boston Harbor Islands is much more complex and inclusive than Edward Rowe Snow's gripping but somewhat one-dimensional tales of adventure and pirate treasure. Endeavoring to educate the public about the many aspects of the islands' geography, ecology, and history, the four major "park

FIGURE 29. The *Edward Rowe Snow* arriving at Long Wharf. Its namesake was instrumental in disseminating the perception of the Boston Harbor Islands as historic, storied places.—*Courtesy of Chris Wood via Wikimedia Commons. This work is licensed for use under Creative Commons license CC BY-SA 4.0*

themes" identified in the park's *General Management Plan* attempt to integrate both the islands' current uses and the less obvious aspects of their past. The park themes, emphasizing the interconnectedness of natural and human history, seek to present the islands' current form and function in the context of their geology, ecology, and past human settlement as well as their historical ties to Boston. The theme "Islands on the Edge" focuses on the geographical and historical position of the harbor islands on the edge of the continent, the city, and metaphorically also on the edge of society. "Home in the Harbor" thematizes both wildlife habitats and human habitation on the harbor islands: Native American seasonal camps as well as the farmers, fishers, and facility keepers and other islanders of modern times. The "Portal to New England" theme showcases the historical roles of the Boston Harbor space as port of entry and the islands' roles as sites of lighthouses and coastal defenses, while the last thematic cluster, "Renewal and Reconnection," focuses on the harbor's recent history of cleanup and ecological preservation and at the same time presents the islands as places where the residents of the metropolitan area can seek relaxation and personal renewal.[8] Despite this complex and nuanced interpretation, the overarching park theme accentuates less the islands' connections and their historical belonging to the city than their separation from Boston and their essential otherness. The park's official motto, "Minutes Away. Worlds Apart," situates the harbor islands physically close to the city, but presents them as different from it in every other respect: their nature, their stories and physical form, the experience of visiting them. To go out to the islands has become synonymous with an out-of-the-ordinary experience, with the possibility to escape the world of the city for a few hours. The currently prevailing narrative constructs the harbor islands as not only different but unique—in their history and nature, in the opportunities they offer for recreation. "When you visit the Boston Harbor Islands," the *Harbor Islands Times* claimed in 2016, "you can step back in time and away from the everyday. There is no place in New England with such a diverse wealth of opportunities for exploration, education, and relaxation."[9] Regarding the harbor islands as a refuge from the dreariness of everyday life became the accepted way of looking at them: public and private initiatives around the harbor as well as popular guidebooks advertise trips to the harbor islands as a way of escaping the routine of life in the city. According to the *Lonely Planet* guidebook, a visit to the islands is a "unique opportunity for outdoor adventure," while other guides point out pirate treasure and ghosts that could be encountered on the harbor islands.[10] A flyer of Charlestown's Constitution Marina from the early 2000s selling mooring spaces at the islands tried to attract customers using a slogan almost identical to the park motto: "Escape to the Boston Harbor Islands: A whole world of difference just minutes from downtown."[11] While most park materials and tourist brochures emphasize the islands' accessibility, at the same time they make clear that the islands' attractiveness rests in the respite they offer and in their difference from the city.

During the almost four hundred years of Boston's existence, its harbor islands have undergone tremendous changes in their ecology, their topography, and their human uses. These transformations have been governed by the changing way Bostonians related to and used the islands. They started out as integral parts of the early colonial town of the seventeenth century, providing pastureland and resources such as timber and hay. The amphibious culture of early New England did not regard the water between the islands and Boston as an interval but rather as a connecting medium. The growing and increasingly terrestrial metropolis of the nineteenth century pushed the islands on the urban fringe and assigned them to uses that would have been only grudgingly tolerated within mainland city limits. Through these unsavory yet necessary uses, however, the city identified the islands as a firm part of its infrastructure. Mid-twentieth-century Boston, casting about desperately for ways to replenish its ailing finances and renew its run-down core, saw the islands as vacant urban space that the city could use to reinvent itself. The latest chapter in the islands' history has cast them in the role of the city's open space: by regarding the islands as romantic and essentially different from the city, yet at the same time shaping them according to the city's recreational needs, Boston has created an urban recreational area, a useful wilderness.

Whether as dumping grounds or as jewels in the crown of the recreational Boston Harbor, the islands' uses have nearly always grown out of their ambivalent position vis-à-vis the city. Perceived as separate units surrounded by water, the islands' apparent detachment from the mainland has allowed for a variety of interpretations over the years. As if attached by a rubber band, the islands were drawn closer and pushed farther away from the city depending on their accessibility, their current uses, and their prevailing perception. Each period gave new meaning to their islandness: what was interpreted in the seventeenth century as advantageous and accessible locations became regarded in the nineteenth and early twentieth centuries as remote and dismal sites at the bottom of the city's hierarchy of places, only to emerge as spots of romantic and desirable isolation in the latter part of the twentieth and early twenty-first centuries.

Today, the Boston Harbor Islands are more present in downtown Boston than ever before. The Boston Harbor Islands Welcome Center in downtown Boston advertises the attractions of the islands with tranquil nature photographs, presenting them as a verdant natural area offering an abundance of vacation-like activities only a short boat ride away. They are also exceptionally accessible: the welcome center is located just across the street from the historic Long Wharf from which public boats carry visitors to the harbor islands every hour from the beginning of May until Columbus Day in early October. At the same time, the islands never seemed to be less part of Boston than today. Their function for the city depends on their image of being nonurban: on their emphasized otherness. Their islandness is being translated as isolation, but a permeable one: an isolation that sets them apart from the city, yet one that can be traversed in only twenty minutes to cater to Bostonians' recreational needs.

The idea of islands as wild and extraordinary places fundamentally different from the mainland has roots in historical European thought about islands as small universes contained in themselves and in the imagination of insular settings as "enchanted isles," a tradition that goes back to the Renaissance and in some respects even to antiquity. However, the notion that we take for granted today, namely, that this claim of uniqueness and otherworldliness applies to any island, including the historically urban Boston Harbor Islands, emerged only in the past one hundred years and became the primary force shaping the islands even later. For the largest part of Boston's history, the islands formed Boston's marine suburb. The idea that their islandness was a quality that set them apart from all other places and made them seats of mystery and adventure is a relatively recent invention. The decisive factor in this sea change in perception was the way in which Bostonians interacted with their harbor and its islands. In the days when Boston was primarily a seaport, the islands were part of a harbor space to which the city's inhabitants related mainly through labor. Colonial Bostonians harvested wood and gravel on the islands and farmed them seasonally. The nineteenth century that placed industries and municipal facilities on the islands reinforced their identity as working landscapes and homes to small populations of laborers, factory workers, facility keepers, and fishermen. It was only with the slow demise of Boston's shipping and the decline of the port that the landscape of the harbor islands became unfamiliar to most city dwellers and open to new interpretations. The current way of relating to the islands through leisure depends on and strengthens their image as places outside everyday experience that provide a sense of adventure and escape from the city.

While the layers of history are mere recollections on many of the harbor islands, on Spectacle Island the past is still materially present. The Big Dig excavate with which the island had been capped obscured its history from view but did not obliterate it completely. In 2007 the artist Teri Rueb was inspired by Spectacle's layered past to create an installation titled *Core Sample*. The centerpiece of the artwork was a sound walk mapping the island's many overlapping layers, proceeding from plantings, top soil, and the Central Artery fill over the landfill and industries Spectacle had once been home to and over the Native American midden all the way down to the twin drumlins' geological core. The "sound sculpture," as Rueb called it, was meant to represent the deposits left by the island's material, cultural, and social history, "questioning what is seen versus what is not seen, what is preserved and recorded versus what is suppressed and denied."[12] Rueb's project is a creative way of approaching the island's complex history that remains hidden and in plain view at the same time. To visitors unfamiliar with Spectacle's history, nothing in the island's appearance hints at its past uses: what meets the eye is green parkland conforming to the Boston Harbor Islands' current image as a recreational area. The island's shape, however, has been determined just as much by its past perceptions: the way Spectacle looks today is the result of the preexisting landfill, which in turn was placed on the island only because of its previous waste-processing uses. Spectacle as it is today, much like New York City's Fresh Kills, is an accidental social and environmental sculpture, a material testament to Bostonians' culture, history, and their changing values.[13]

The layers of history that are so conspicuously present on Spectacle can be detected in more subtle forms on all the other Boston Harbor Islands as well. Every island in this urban archipelago bears traces of its past human uses. Islands like Nixes Mate or Slate Island in Hingham Bay are now a mere shadow of their former selves, depleted by decades of gravel quarrying and subsequent erosion. The sewage-treatment plant on Deer Island, the causeways that attach Nut and Moon Islands to the mainland, and the hospital complex on Long Island are all products of the nineteenth-century notion of the harbor islands as Boston's dumping ground for unwanted materials and people. The preserved fortress on Georges Island and the wild and rugged look of the Brewsters go back to the late-twentieth-century perception of the harbor islands as places fit for recreation and as sites of historicity and natural beauty. Although created by the geological forces of the last ice age, the islands have been in an almost equal measure shaped by the city at the head of the bay.

The history of the Boston Harbor Islands holds important lessons about urban islands in general. Dozens of islands near coastal cities all across North America went through very similar development, suggesting a common pattern in the perception and use of urban islands. The islands of New York Harbor, especially those of Jamaica Bay, were historically used for very much the same purposes as Boston's harbor islands, providing space for quarantine hospitals, landfills, and airports; since 1972 they have been united in the Gateway National Recreation Area, offering swimming, biking, and hiking opportunities to the population of the metropolis.[14] Baltimore has joined three small islands in its harbor, Hart, Miller, and Pleasant Islands, with a dike in the 1980s and, after filling the area with material dredged from its harbor and local marinas, declared the resulting Hart-Miller Island a state park.[15] The islands of the Sacramento–San Joaquin River delta were used extensively for agriculture in the nineteenth century only to be turned into state parks in the 1950s.[16] Montreal turned its river islands into exhibition grounds, while Toronto developed its lakeshore islands into pleasure grounds for the city.[17] These developments suggest a widespread perception of urban islands as unique places fit for very specific uses; places that are at the same time attached to and detached from the bodies of "their" cities.

Those who visit Spectacle Island today find that its beaches still glitter in the sun. They are strewn with sea glass, small shards of broken bottles and pottery that escaped the landfill and washed back ashore, ground and smoothed by the sea. In the 1960s, the sea glass served as evidence that the island was a dismal wasteland: the 1969 harbor documentary featured footage of Spectacle's glimmering shores as an argument in a story of the harbor islands' misuse and dilapidation.[18] Today, the same sea glass is one of Spectacle's most treasured features: it is advertised by the park management as the island's particular attraction, and tourists are cautioned not to take any off the island, so that future visitors can enjoy it as well. Like the sea glass, the way the Boston Harbor Islands have been treated in the course of Boston's history has depended on Bostonians' changing perceptions of their city's insular appendages. Although the way they have been regarded has alternated between belonging and otherness, and although their uses included such disparate items as poorhouses and parks, for the past four hundred years the harbor islands have never ceased to be part of Boston. Despite appearances, the Boston Harbor Islands are as urban as Boston Common. When you stand today on Boston's waterfront and look across the bay to the islands, they seem peaceful and distant, like another world compared to the noisy, restless metropolis behind you. Yet what you are looking at is part of Boston, its urban archipelago: places that have over centuries been shaped by the city's perceptions and uses. When Boston looks across the water, it sees itself in the islands: they are a mirror of the city's own ever-changing features.

NOTES

Introduction

1. The only islands in Boston Harbor that are not part of the Boston Harbor Islands National and State Park are Castle Island, a former island that is now part of the peninsula of South Boston, and Spinnaker (Hog) Island in Hull, now occupied by a condominium.
2. Chris Sweeney, "The Guide: Boston Harbor Islands," *Boston Magazine*, May 1, 2016; Thomas Oide, "Magic, Vibe of the Boston Harbor Islands Lingers," *Boston Globe*, July 16, 2018; Valerie A. Russo, "Shore Thing: Online Map Reveals Bay State Coastal Gems," *Patriot Ledger*, August 20, 2018; "Summer as a Verb," *Improper Bostonian*, May 20, 2016. For guidebooks, see Fred Bouchard, *Boston*, 28th ed. (New York: Fodor's Travels, 2014); Barbara Rockwell, *Boston*, 5th ed. (Singapore: APA, 2013); and Mara Vorhees, *Boston*, 4th ed. (London: Lonely Planet, 2009).
3. Boston was indeed by far not the only American city to alter its environment in a radical way. As Michael Rawson remarks in his *Eden on the Charles: The Making of Boston* (Cambridge, MA: Harvard University Press, 2014), 1, such interventions in the cities' natural settings often accompanied nineteenth-century urbanization. For instance, Chicago reversed the flow of its river, in Pittsburgh parts of the city were raised by almost ten feet to improve drainage, and New Orleans built an extensive system of levees and dikes to contain the Mississippi River and Lake Pontchartrain.
4. Nancy S. Seasholes, *Gaining Ground: A History of Landmaking in Boston* (Cambridge, MA: MIT Press, 2003), 2.
5. The literature about the topographical development and geographical expansion of Boston is vast. An authoritative introduction to Boston's physical evolution from its founding until the late twentieth century is Walter Muir Whitehill and Lawrence W. Kennedy, *Boston: A Topographical History*, 3rd ed. (Cambridge, MA: Belknap Press of Harvard University Press, 2000). Even general histories such as Thomas H. O'Connor, *The Hub: Boston Past and Present* (Boston: Northeastern University Press, 2001) prominently feature the city's geographical development. The growth of colonial Boston is the subject of Darrett B. Rutman, *Winthrop's Boston: Portrait of a Puritan Town, 1630–1649* (Chapel Hill: University of North Carolina Press, 1965), as well as the classic study of early American towns by Carl Bridenbaugh, *Cities in the Wilderness:*

The First Century of Urban Life in America, 1625–1742 (New York: Ronald Press, 1938). Boston city planning is the subject of Lawrence W. Kennedy, *Planning the City upon a Hill: Boston since 1630* (Amherst: University of Massachusetts Press, 1992); Thomas H. O'Connor, *Building a New Boston: Politics and Urban Renewal, 1950–1970* (Boston: Northeastern University Press, 1993); and Anne Whiston Spirn, *The Granite Garden: Urban Nature and Human Design* (New York: Basic Books, 1984). A number of studies focus on Boston's suburbs, most prominently Sam Bass Warner, *Streetcar Suburbs: The Process of Growth in Boston, 1870–1900* (Cambridge, MA: MIT Press, 1962); and James C. O'Connell, *The Hub's Metropolis: Greater Boston's Development from Railroad Suburbs to Smart Growth* (Cambridge, MA: MIT Press, 2013). Jane Holtz Kay, *Lost Boston* (Boston: Houghton Mifflin, 1980), offers an overview of Boston's architectural history. Boston's celebrated park system is the subject of Cynthia Zaitzevsky, *Frederick Law Olmsted and the Boston Park System* (Cambridge, MA: Belknap Press of Harvard University Press, 1982), while another study focuses solely on the transformation of the Charles River estuary into a water park: Karl Haglund, *Inventing the Charles River* (Cambridge, MA: MIT Press, 2003). Recent years saw the publication of two outstanding volumes on Boston's environmental history: Anthony N. Penna and Conrad E. Wright, eds., *Remaking Boston: An Environmental History of the City and Its Surroundings* (Pittsburgh: University of Pittsburgh Press, 2009); and Rawson, *Eden on the Charles*.

6. Boston Harbor and its history have only recently begun to receive historians' attention. While the history of Boston's maritime commerce and its famous shipbuilding was popular with earlier generations of historians and featured prominently in the writings of such eminent Massachusetts historians as Samuel Eliot Morison, in *The Maritime History of Massachusetts, 1783–1860* (Boston: Houghton Mifflin, 1921), the history of Boston Harbor not just as a port but as a place in its own right and as a natural environment has only recently come into the limelight. The principal publications concerned with the environmental history of Boston Harbor are Penna and Wright, *Remaking Boston*; and Rawson, *Eden on the Charles*. Michael Rawson is also the author of an article probing the nineteenth-century debate about a perceived shoaling of the harbor, "What Lies Beneath: Science, Nature, and the Making of Boston Harbor," *Journal of Urban History* 35, no. 5 (2009). The most detailed history of Boston shaping its shoreline is Seasholes, *Gaining Ground*. There are a number of books and articles focusing on the political and legal history of the Boston Harbor cleanup in the 1980s and 1990s: the principal studies are Charles M. Haar, *Mastering Boston Harbor: Courts, Dolphins, and Imperiled Waters* (Cambridge, MA: Harvard University Press, 2005); and Eric Jay Dolin, *Political Waters: The Long, Dirty, Contentious, Incredibly Expensive but Eventually Triumphant History of Boston Harbor—a Unique Environmental Success Story* (Amherst: University of Massachusetts Press, 2004).
7. As of now, there is no scholarly historical study of the Boston Harbor Islands. There are a number of nineteenth-century guidebooks that offer descriptions of the islands' geography and histories. See, for instance, Nathaniel B. Shurtleff, *A Topographical and Historical Description of Boston* (Boston: City Council, 1871); James H. Stark, *Illustrated History of Boston Harbor* (Boston: Photo-Electrotype, 1879); and M. F. Sweetser, *King's Handbook of Boston Harbor*, 3rd ed. (Boston: Moses King, 1889). Most contemporary popular books about the islands focus on colorful anecdotes from the islands' history and on their recreational potential, such as David Kales, *Boston Harbor Islands: A History of an Urban Wilderness* (Charleston, SC: History Press, 2007), or Christopher Klein, *Discovering the Boston Harbor Islands: A Guide to the City's Hidden Shores*, 2nd ed. (Boston: Union Park Press, 2011). A popular yet well-researched and informative history of the

Boston Harbor Islands is Stephanie Schorow, *East of Boston: Notes from the Harbor Islands* (Charleston, SC: History Press, 2008). The most prolific author of popular books about the Boston Harbor Islands' history was Edward Rowe Snow, whose *The Islands of Boston Harbor* appeared in three editions, in 1935, 1971, and 2002. See Edward Rowe Snow, *The Islands of Boston Harbor: Their History and Romance, 1626–1935* (Andover, MA: Andover Press, 1935), *The Islands of Boston Harbor, 1630–1971*, 2nd ed. (New York: Dodd, Mead, 1971), and *The Islands of Boston Harbor*, with the assistance of Jeremy D'Entremont (Carlisle, MA: Applewood Books, 2002). There are a number of popular and scientific publications focusing on the flora of the harbor islands, including Dale F. Levering, *An Illustrated Flora of the Boston Harbor Islands* (Boston: Northeastern University Press, 1978), or a special issue of the *Northeastern Naturalist* devoted to the Boston Harbor Islands National Park Area: Charles T. Roman, Bruce Jacobson, and Jack Wiggin, eds., "Boston Harbor Islands National Park Area: Natural Resources Overview," special issue, *Northeastern Naturalist* 12, no. 3 (2005). Historical studies about Boston, if they mention the harbor islands at all, mostly do so only in passing. Michael Rawson makes several mentions of them in connection with the perceived shoaling of Boston Harbor in the mid-nineteenth century in *Eden on the Charles* and "What Lies Beneath." William B. Meyer lists the islands as the urban fringe of late-eighteenth-century Boston in "A City (Only Partly) on a Hill: Terrain and Land Use in Pre-Twentieth-Century Boston," in *Remaking Boston*, ed. Penna and Wright, 127–47.

8. The term *urban islands* has been used by island studies scholars, especially Adam Grydehøj. See Grydehøj, "Understanding Island Cities," *Island Studies Journal* 9, no. 2 (2014); Grydehøj et al., "Returning from the Horizon: Introducing Urban Island Studies," *Urban Island Studies* 1, no. 1 (2015); and Grydehøj and Marco Casagrande, "Islands of Connectivity: Archipelago Relationality and Transport Infrastructure in Venice Lagoon," *Area* 52, no. 1 (2020), as well as a chapter on islands and cities in Godfrey Baldacchino, ed., *The Routledge International Handbook of Island Studies: A World of Islands* (London: Routledge, 2018). Their definition of an urban island, however, is much broader, including both heavily urbanized archipelagos, such as Venice, and major population centers of larger islands, such as Reykjavík. I focus here on a specific type of urban islands close to coastal cities that, while not necessarily obviously urbanized, have nevertheless shared close historical ties with "their" cities.
9. In a very broad sense, the Boston Harbor Islands could be described as heterotopias, a concept proposed by Michel Foucault in his 1967 lecture titled "Des espaces autres." In it, Foucault suggests a category of spaces that invert the nature of other spaces, while at the same time reflecting and representing them. However, the notion of heterotopias appears far too broad and generalizing to be useful as an analytical tool in this particular case. See Foucault, "Of Other Spaces: Utopias and Heterotopias," *Architecture/Mouvement/Continuité* (October 1984).
10. There exist several testimonies about life on Spectacle Island: onetime islanders Elizabeth Wyatt-Hargreaves, Robert Wyatt, and Kathleen Rivera have all written memoirs or given interviews recalling their time on Spectacle Island. See Billie Wyatt-Hargreaves, diary, transcribed by Jeff Peters, June 2004, Spectacle Island Interpretive Guide, City of Boston Archives; Robert H. Wyatt, "A Short Description of Spectacle Island and What Transpired There as Remembered by Me" (unpublished manuscript, 2001), Suzanne Gall Marsh Collection, University Archives and Special Collections, Joseph P. Healey Library, University of Massachusetts, Boston; and Kathleen A. Rivera, interview by Boston Harbor Islands Mass. Memories Road Show, September 17, 2011, University Archives and Special Collections, Joseph P. Healey Library, University of Massachusetts, Boston.

11. Laura Alice Watt, *The Paradox of Preservation: Wilderness and Working Landscapes at Point Reyes National Seashore* (Berkeley: University of California Press, 2017); James W. Feldman, *A Storied Wilderness: Rewilding the Apostle Islands* (Seattle: University of Washington Press, 2011).
12. Carey Goldberg, "Emerald Isles, Boston's Own," *New York Times*, July 14, 2000.
13. For a discussion of the relative positions of islands vis-à-vis continents in the historical European imagination, see John R. Gillis, *Islands of the Mind: How the Human Imagination Created the Atlantic World* (New York: Palgrave Macmillan, 2004) and *The Human Shore: Seacoasts in History* (Chicago: University of Chicago Press, 2012); and Martin W. Lewis and Kären Wigen, *The Myth of Continents: A Critique of Metageography* (Berkeley: University of California Press, 1997).
14. Christopher L. Pastore, *Between Land and Sea: The Atlantic Coast and the Transformation of New England* (Cambridge, MA: Harvard University Press, 2014); Matthew G. McKenzie, *Clearing the Coastline: The Nineteenth-Century Ecological & Cultural Transformation of Cape Cod* (Hanover, NH: University Press of New England, 2010); John T. Cumbler, *Cape Cod: An Environmental History of a Fragile Ecosystem* (Amherst: University of Massachusetts Press, 2014); Gillis, *Human Shore*; John R. Stilgoe, *Alongshore* (New Haven, CT: Yale University Press, 1994).
15. Prominent examples include Matthew W. Klingle, *Emerald City: An Environmental History of Seattle* (New Haven, CT: Yale University Press, 2007); Matthew Morse Booker, *Down by the Bay: San Francisco's History between the Tides* (Berkeley: University of California Press, 2013); Stéphane Castonguay and Michèle Dagenais, eds., *Metropolitan Natures: Environmental Histories of Montreal* (Pittsburgh: University of Pittsburgh Press, 2011); Michèle Dagenais, *Montreal, City of Water: An Environmental History* (Vancouver: UBC Press, 2017); Ari Kelman, *A River and Its City: The Nature of Landscape in New Orleans* (Berkeley: University of California Press, 2003); Craig E. Colten, *An Unnatural Metropolis: Wresting New Orleans from Nature* (Baton Rouge: Louisiana State University Press, 2006); and Eric W. Sanderson, *Mannahatta: A Natural History of New York City* (New York: Abrams, 2009).
16. Lewis and Wigen, *Myth of Continents*.
17. Gillis, *Islands of the Mind*; Gillis, *Human Shore*; John R. Gillis, "Islands in the Making of Atlantic Oceania, 1500–1800," in *Seascapes: Maritime Histories, Littoral Cultures, and Transoceanic Exchanges*, ed. Jerry H. Bentley, Renate Bridenthal, and Kären Wigen (Honolulu: University of Hawai'i Press, 2007).
18. While John Gillis's *Human Shore* and *Islands of the Mind* are excellent studies of the place of islands and the coast in human history, their focus is too broad to deal with urban islands in greater detail. Pastore, *Between Land and Sea*, which deals with the littoral ecology of New England from the seventeenth to the nineteenth centuries, mentions the islands of Narragansett Bay only in passing and does not treat them as a separate category. Other environmental histories of New England, such as Blake A. Harrison and Richard W. Judd, eds., *A Landscape History of New England* (Cambridge, MA: MIT Press, 2011), or Richard W. Judd, *Second Nature: An Environmental History of New England* (Amherst: University of Massachusetts Press, 2014), are similarly too broad in their focus on the entire region to devote special attention to coastal islands. Environmental histories of coastal cities tend to mention their coastal and river islands mostly as an afterthought; they never take center stage (for examples, see note 15). The emerging field of blue humanities, represented by such volumes as Helen M. Rozwadowski, *Fathoming the Ocean: The Discovery and Exploration of the Deep Sea* (Cambridge, MA: Belknap Press of Harvard University Press, 2008) and *Vast Expanses: A History of the Oceans* (London: Reaktion Books, 2018), or David Armitage, Alison Bashford, and Sujit Sivasundaram,

eds., *Oceanic Histories* (Cambridge: Cambridge University Press, 2018), focusing on the human history of the ocean, usually finds coastal islands to be too terrestrial and gives them little attention. One of the few books that see the ocean as an amphibious space and focus on its edges as well as its depths is John R. Gillis and Franziska Torma, eds., *Fluid Frontiers: New Currents in Marine Environmental History* (Cambridge: White Horse Press, 2015). Studies focusing solely on islands tend to direct their gaze toward larger and more remote islands and deal with topics of resilience, climate change, and environmental hazards, while leaving coastal and urban islands out of focus. See Rebecca Hofmann and Uwe Lübken, eds., "Small Islands and Natural Hazards," special issue, *Global Environment* 8, no. 1 (2015); Patrick D. Nunn, *Vanished Islands and Hidden Continents of the Pacific* (Honolulu: University of Hawai'i Press, 2009); John Connell, *Islands at Risk? Environments, Economies and Contemporary Change* (Cheltenham: Edward Elgar, 2013); and Toyin Falola, R. J. Parrott, and Danielle P. Sanchez, eds., *African Islands: Leading Edges of Empire and Globalization* (Rochester, NY: University of Rochester Press, 2019). An overwhelming majority of the articles published in the *Island Studies Journal* focus on islands that are relatively far from the mainland and are home to human communities or even form independent states; typical examples include the Azores, the Canary Islands, or the Maldives. In recent years, there has been a rise of interest in urban islands within the field of island studies, as evidenced especially by the publications of Adam Grydehøj (see note 8) and the articles published in the now-defunct journal *Urban Island Studies*. A recent historical study of an urban island with a development trajectory bafflingly similar to Spectacle Island's is Miriam Sicherman, *Brooklyn's Barren Island: A Forgotten History* (Charleston, SC: History Press, 2019).

19. The periodization of the islands' development I use here roughly corresponds with two of the regimes of resource utilization John T. Cumbler introduces in his book on the environmental history of Cape Cod: the transition from a regime of resource extraction and production to one of distant resources and a local economy dependent on recreation and tourism. However, while Cumbler's main interest is the impact of these regimes on Cape Cod's ecology, these divisions do not form a core part of my narrative. See Cumbler, *Cape Cod*.

Chapter 1: The Commons of Boston Harbor

1. Richard Mather, "Richard Mather's Journal," in *Chronicles of the First Planters of the Colony of Massachusetts Bay, from 1623 to 1636*, ed. Alexander Young (Boston: Charles C. Little and James Brown, 1846); B. R. Burg, *Richard Mather of Dorchester* (Lexington: University Press of Kentucky, 1976).
2. For detailed accounts of the Atlantic crossing from the 1630s, see, for instance, Francis Higginson, "A True Relation of the Last Voyage to New-England," in *Chronicles of the First Planters*, ed. Young; and James K. Hosmer, ed., *Winthrop's Journal: History of New England, 1630–1649*, vol. 1 (New York: Charles Scribner's Sons, 1908).
3. Higginson, "True Relation of the Last Voyage to New-England," 226. What seventeenth-century authors referred to as "grampus" was probably one of the large oceanic dolphin species, either the long-finned pilot whale (*Globicephala melas*) or Risso's dolphin (*Grampus griseus*).
4. Mather, "Richard Mather's Journal," 471. For more information on European fishing colonies along the coast of New England and Canada, see W. Jeffrey Bolster, "Putting the Ocean in Atlantic History: Maritime Communities and Marine Ecology in the Northwest Atlantic, 1500–1800," *American Historical Review* 113, no. 1 (2008).

5. "Deputy Governor Thomas Dudley's Letter to the Countess of Lincoln, 12 March 1631," in *Chronicles of the First Planters*, ed. Young, 312.
6. In the early days of the colony, both Boston Harbor and Massachusetts Bay between Cape Ann and Cape Cod were known as the Bay of Massachusetts. Boston Harbor would be called the "still bay" to distinguish it from the larger and not-so-still outer Bay of Massachusetts. See William Wood, *New England's Prospect: A True, Lively, and Experimental Description of That Part of America, Commonly Called New England* (London: Thomas Cotes and John Bellamie, 1634), 3.
7. Hosmer, *Winthrop's Journal*, 50.
8. Despite appearances, the Graves did not derive their somber name from being the graveyard of ships and sailors—although shipwrecks happened at them often enough. Rather, they were named for Thomas Graves (1605–53), a prominent personality of early colonial Massachusetts.
9. William F. Poole, ed., *Edward Johnson: Wonder-Working Providence: Of Sions Saviour in New England*, vol. 2 (Andover, MA: W. F. Draper, 1867), 37.
10. Hosmer, *Winthrop's Journal*; "Anthony Thacher's Narrative of His Shipwreck," in *Chronicles of the First Planters*, ed. Young; John R. Stilgoe, "A New England Coastal Wilderness," *Geographical Review* 71, no. 1 (1981).
11. William Cronon, *Changes in the Land: Indians, Colonists, and the Ecology of New England*, 1st rev. ed. (New York: Hill and Wang, 2003); Richard W. Judd, *Second Nature: An Environmental History of New England* (Amherst: University of Massachusetts Press, 2014); Julie A. Richburg and William A. Patterson, "Historical Description of the Vegetation of the Boston Harbor Islands, 1600–2000," *Northeastern Naturalist* 12, no. 3 (2005).
12. Hosmer, *Winthrop's Journal*, 105.
13. Wood, *New England's Prospect*, 42.
14. Petra J. E. M. van Dam, "De amfibische cultuur: Een visie op watersnoodrampen," talk given at the Vrije Universiteit Amsterdam, October 29, 2010; John T. Cumbler, *Cape Cod: An Environmental History of a Fragile Ecosystem* (Amherst: University of Massachusetts Press, 2014), 41. The adaptation of early Bostonians to their watery environment is also the topic of Christopher L. Pastore, "Filling Boston Commons: Law, Culture, and Ecology in a Seventeenth-Century Estuary," in *Fluid Frontiers: New Currents in Marine Environmental History*, ed. John R. Gillis and Franziska Torma (Cambridge: White Horse Press, 2015), 27–38.
15. Hosmer, *Winthrop's Journal*, 90.
16. *Records of the Court of Assistants of the Colony of the Massachusetts Bay, 1630–1692*, vol. 2 (Boston: County of Suffolk, 1904); Young, *Chronicles of the First Planters*, 323.
17. Hosmer, *Winthrop's Journal*, 55.
18. John Josselyn, *An Account of Two Voyages to New-England, Made during the Years 1638, 1663* (Boston: William Veazie, 1865), 20.
19. Judd, *Second Nature*.
20. Sherburne Friend Cook, *The Indian Population of New England in the Seventeenth Century* (Berkeley: University of California Press, 1976); Cronon, *Changes in the Land*; Judd, *Second Nature*.
21. James D. Drake, *King Philip's War: Civil War in New England, 1675–1676* (Amherst: University of Massachusetts Press, 1999).
22. On the social makeup and motivations for migration of the early Massachusetts settlers, see David Grayson Allen, *In English Ways: The Movement of Societies and the Transferal of English Local Law and Custom to Massachusetts Bay in the Seventeenth Century* (New York: W. W. Norton, 1982).

23. See Epeli Hau'ofa, "Our Sea of Islands," *Contemporary Pacific* 6, no. 1 (1994); John R. Gillis, "Islands in the Making of Atlantic Oceania, 1500–1800," in *Seascapes: Maritime Histories, Littoral Cultures, and Transoceanic Exchanges*, ed. Jerry H. Bentley, Renate Bridenthal, and Kären Wigen (Honolulu: University of Hawai'i Press, 2007). The term *ocean highway* is used by Carl Bridenbaugh in his classic study of early urban America *Cities in the Wilderness: The First Century of Urban Life in America, 1625–1742* (New York: Ronald Press, 1938), 5.
24. Josselyn, *Account of Two Voyages to New-England*, 10–11; Paul J. Lindholdt, introduction to *John Josselyn, Colonial Traveler: A Critical Edition of Two Voyages to New-England* (Hanover, NH: University Press of New England, 1988), xvi.
25. Gillis, "Islands in the Making of Atlantic Oceania," 24; John R. Gillis, *The Human Shore: Seacoasts in History* (Chicago: University of Chicago Press, 2012), 54. For a discussion of how the term *continent* and the Europeans' understanding of their world's geography evolved since antiquity, see Martin W. Lewis and Kären Wigen, *The Myth of Continents: A Critique of Metageography* (Berkeley: University of California Press, 1997), 21–46.
26. Higginson, "True Relation of the Last Voyage to New-England," 232, 234.
27. Darrett B. Rutman, *Winthrop's Boston: Portrait of a Puritan Town, 1630–1649* (Chapel Hill: University of North Carolina Press, 1965), 35.
28. Wood, *New England's Prospect*, 42.
29. Bridenbaugh, *Cities in the Wilderness*, 9.
30. Dale F. Levering, "The Changing Flora of the Boston Harbor Islands," *Arnoldia* 43, no. 3 (1988), 18.
31. Alexander Young, ed., *Chronicles of the Pilgrim Fathers of the Colony of Plymouth, from 1602 to 1625* (Boston: Charles C. Little and James Brown, 1841), 229.
32. Boston archaeologist Barbara E. Luedtke has published widely on the Native archaeology of the Boston Harbor Islands. See Barbara E. Luedtke and Peter S. Rosen, "Archaeological Geology of Long Island, Boston Harbor," in *Field Trip Guidebook for the Northeastern United States* (Boston: Geological Society of America, 1993); Luedtke, *The Archaeology of Thompson Island* (Boston: Department of Archaeology, University of Massachusetts, 1996); and Luedtke, "Archaeology on the Boston Harbor Islands after 25 Years," *Bulletin of the Massachusetts Archaeological Society* 61, no. 1 (2000).
33. Rutman, *Winthrop's Boston*, 39.
34. *Records of the Court of Assistants*, 16–17.
35. Michael Rawson, *Eden on the Charles: The Making of Boston* (Cambridge, MA: Harvard University Press, 2014), 25; Allen, *In English Ways*, 30–33, 46–47.
36. Cumbler, *Cape Cod*, 33.
37. *Records of the Court of Assistants*, 22, 31.
38. See Rutman, *Winthrop's Boston*.
39. *Records of the Court of Assistants*, 14.
40. Hosmer, *Winthrop's Journal*, 258.
41. William A. Stokinger, "Spectacle Island Chronology Draft" (unpublished manuscript, May 2008), Massachusetts Department of Conservation and Recreation, Division of Urban Parks and Recreation, Bureau of Ranger Services; Rutman, *Winthrop's Boston*, 85.
42. Rutman, *Winthrop's Boston*, 84–85.
43. Rawson, *Eden on the Charles*; Nathaniel B. Shurtleff, *A Topographical and Historical Description of Boston* (Boston: City Council, 1871), 304; Rutman, *Winthrop's Boston*, 92. For more detail on the agricultural character of seventeenth-century Boston, see David B. Landon, ed., "Feeding Colonial Boston: A Zooarchaeological Study," special issue, *Historical Archaeology* 30, no. 1 (1996).

44. Christopher L. Pastore, *Between Land and Sea: The Atlantic Coast and the Transformation of New England* (Cambridge, MA: Harvard University Press, 2014), 22, 58, 59, 147–48, 161.
45. Albert B. Hart, ed., *Hamilton's Itinerarium: Being a Narrative of a Journey from Annapolis, Maryland, through Delaware, Pennsylvania, New York, New Jersey, Connecticut, Rhode Island, Massachusetts and New Hampshire from May to September 1744, by Doctor Alexander Hamilton* (St. Louis: William K. Bixby, 1907), 59; William Bentley, *The Diary of William Bentley, D.D., Pastor of the East Church, Salem, Massachusetts* (Salem, MA: Essex Institute, 1905), 284; Philip W. Conkling, *Islands in Time: A Natural and Cultural History of the Islands of the Gulf of Maine*, 2nd ed. (Rockland, ME: Island Institute / Down East Books, 1999), 105, 110.
46. Mark Peterson, *The City-State of Boston: The Rise and Fall of an Atlantic Power, 1630–1865* (Princeton, NJ: Princeton University Press, 2019), 35; Thomas H. O'Connor, *The Hub: Boston Past and Present* (Boston: Northeastern University Press, 2001), 5–11.
47. Wood, *New England's Prospect*, 42.
48. Young, *Chronicles of the Pilgrim Fathers*; Thomas Prince, *A Chronological History of New England in the Form of Annals* (Boston: Kneeland & Green, 1736), 198.
49. For a detailed discussion of Boston Harbor geology, see Peter S. Rosen and Duncan M. FitzGerald, "The Drowning of Boston Harbor and the Development of the Shoreline," in *Remaking Boston: An Environmental History of the City and Its Surroundings*, ed. Anthony N. Penna and Conrad E. Wright (Pittsburgh: University of Pittsburgh Press, 2009), 19–32; and Clifford Alan Kaye, *The Geology and Early History of the Boston Area of Massachusetts, a Bicentennial Approach*, Geological Survey Bulletin 1476 (Washington, DC: US Government Printing Office, 1976).
50. Wood, *New England's Prospect*, 3.
51. Mather, "Richard Mather's Journal," 476.
52. Poole, *Edward Johnson*, 63, 194.
53. For a detailed history of the fort on Castle Island, see Shurtleff, *Topographical and Historical Description of Boston*, 478–98.
54. Cumbler, *Cape Cod*, 40.
55. Rutman, *Winthrop's Boston*, 181; Stephen Innes, *Creating the Commonwealth: The Economic Culture of Puritan New England* (New York: W. W. Norton, 1995), 21–22.
56. Hosmer, *Winthrop's Journal*, 127, 65, 109.
57. Wood, *New England's Prospect*; Hosmer, *Winthrop's Journal*, 112.
58. Bolster, "Putting the Ocean in Atlantic History," 26; John J. McCusker and Russell R. Menard, *The Economy of British America, 1607–1789* (Chapel Hill: University of North Carolina Press, 1991), 99.
59. E. H. Derby, *Boston: A Commercial Metropolis: Her Growth, Population, Wealth and Prospects* (Boston: Redding, 1850), 3; Louis E. Cellineri, *Seaport Dynamics: A Regional Perspective* (Lexington, MA: Lexington Books, 1976), 1; David D. Hall, *The Puritans: A Transatlantic History* (Princeton, NJ: Princeton University Press, 2019), 329.
60. The Caribbean islands' political and economic situation was naturally much more complex than it appeared from the perspective of the North American colonies, let alone from the gross simplification presented here. For a more detailed discussion of the early modern Caribbean economy and ecology, see John R. McNeill, *Mosquito Empires: Ecology and War in the Greater Caribbean, 1620–1914* (Cambridge: Cambridge University Press, 2010), 15–62.
61. Rutman, *Winthrop's Boston*; Cellineri, *Seaport Dynamics*, 2.
62. Wendy Warren, *New England Bound: Slavery and Colonization in Early America* (New York: Liveright, 2016) and "'The Cause of Her Grief': The Rape of a Slave in Early New England," *Journal of American History* 93, no. 4 (2007).

63. Cellineri, *Seaport Dynamics*, 3.
64. "Colonial Ordinances of 1641–1647," Massachusetts Court System Law Library, https://www.mass.gov/files/documents/2016/08/qm/colonialordinancesof1651.pdf. One hundred rods equals 1,650 feet, or 500 meters. The law extending waterfront property rights down to the low-water mark is still in place in Massachusetts today, severely limiting public beach access and creating headaches for the state's coastal zone management.
65. Rutman, *Winthrop's Boston*; Josselyn, *Account of Two Voyages to New-England*, 125.
66. Joseph F. Balicki, "Wharves, Privies, and the Pewterer: Two Colonial Period Sites on the Shawmut Peninsula, Boston," *Historical Archaeology* 32, no. 3 (1998): 100.
67. Nancy S. Seasholes, "Filling Boston's Mill Pond," *Historical Archaeology* 32, no. 3 (1998): 121.
68. Walter Muir Whitehill and Lawrence W. Kennedy, *Boston: A Topographical History*, 3rd ed. (Cambridge, MA: Belknap Press of Harvard University Press, 2000), 18.
69. Rutman, *Winthrop's Boston*, 200.
70. Wood, *New England's Prospect*, 3.
71. Balicki, "Wharves, Privies, and the Pewterer"; Seasholes, "Filling Boston's Mill Pond," 14.
72. K. O. Emery et al., "European Cretaceous Flints on the Coast of North America," *Science* 160, no. 3833 (1968); Balicki, "Wharves, Privies, and the Pewterer," 109.
73. Michael Rawson, "What Lies Beneath: Science, Nature, and the Making of Boston Harbor," *Journal of Urban History* 35, no. 5 (2009): 679.
74. Rawson, *Eden on the Charles*, 189–90.
75. Donald G. Jones, *Intensive Archaeological Survey of Spectacle Island in Boston Harbor*, with the assistance of Nancy S. Seasholes (Boston: Office of Public Archaeology, Boston University, 1989), 11.
76. *American Federalist Columbian Centinel*, June 22, 1822.
77. Brian Donahue, "Remaking Boston, Remaking Massachusetts," in *Remaking Boston*, ed. Penna and Wright, 105.
78. Cellineri, *Seaport Dynamics*, 5–6.
79. Lawrence W. Kennedy, *Planning the City upon a Hill: Boston since 1630* (Amherst: University of Massachusetts Press, 1992), 17.
80. Cellineri, *Seaport Dynamics*, 6; Innes, *Creating the Commonwealth*, 288–93.
81. Whitehill and Kennedy, *Boston*, 20.
82. A contemporary description of Long Wharf can be found, for instance, in the travel journal of Alexander Hamilton, a medical doctor from Annapolis, who visited Boston in 1744. See Hart, *Hamilton's Itinerarium*, 131.
83. Whitehill and Kennedy, *Boston*, 21.
84. Yi-fu Tuan, *Space and Place: The Perspective of Experience* (Minneapolis: University of Minnesota Press, 2011), 34 ff.
85. Jacqueline Barbara Carr, *After the Siege: A Social History of Boston, 1775–1800* (Boston: Northeastern University Press, 2005), 48.
86. Pastore, *Between Land and Sea*, 24–26.
87. Carr, *After the Siege*, 80, 68; William B. Meyer, "A City (Only Partly) on a Hill: Terrain and Land Use in Pre-Twentieth-Century Boston," in *Remaking Boston*, ed. Penna and Wright, 128.
88. Rawson, *Eden on the Charles*, 22–29; Carr, *After the Siege*, 68.
89. Carr, *After the Siege*, 82.

Chapter 2: The Ultimate Sink

1. Karl Haglund, *Inventing the Charles River* (Cambridge, MA: MIT Press, 2003), 107.
2. Board of Health of the City of Boston, *Second Annual Report: City Document No. 63* (Boston, 1874), 42–48. The olfactory aspect of the Millers River nuisance is analyzed in detail in Melanie A. Kiechle, *Smell Detectives: An Olfactory History of Nineteenth-Century Urban America* (Seattle: University of Washington Press, 2017), 198–216.
3. Board of Health of the City of Boston, *Sixth Annual Report: City Document No. 68. Documents of the City of Boston for the Year 1878* (Boston, 1879), 3.
4. On the topic of the small olfactory vocabulary, see Kiechle, *Smell Detectives*, 7–8.
5. This list is a compilation of adjectives used to describe odors in the annual reports of the Boston Board of Health. The odors are referred to as vile, offensive, intolerable, dreadful, alarming, and unwholesome in Board of Health of the City of Boston, *Second Annual Report*; as noxious in Board of Health of the City of Boston, *Fourth Annual Report: City Document No. 53. Documents of the City of Boston for the Year 1876* (Boston: Rockwell and Churchill, City Printers, 1877); as bad, foul, and offensive in Board of Health of the City of Boston, *Sixth Annual Report;* as noisome and injurious in Board of Health of the City of Boston, *Twelfth Annual Report: City Document No. 86. Documents of the City of Boston for the Year 1884* (Boston: Rockwell and Churchill, City Printers, 1885); as offensive and objectionable in Health Department of the City of Boston, *Annual Report for the Year 1899–1900: City Document No. 16. Documents of the City of Boston for the Year 1900* (Boston: Municipal Printing Office, 1901); and as offensive and disagreeable in Health Department of the City of Boston, *Annual Report for the Year 1908–1909: City Document No. 18. Documents of the City of Boston for the Year 1909* (Boston: Printing Department, 1910).
6. Stanley K. Schultz, *Constructing Urban Culture: American Cities and City Planning, 1800–1920* (Philadelphia: Temple University Press, 1989), 153; Martin V. Melosi, *The Sanitary City: Environmental Services in Urban America from Colonial Times to the Present*, abridged ed. (Pittsburgh: University of Pittsburgh Press, 2008), 5.
7. For a detailed account of the history of land making in South Boston, see Nancy S. Seasholes, *Gaining Ground: A History of Landmaking in Boston* (Cambridge, MA: MIT Press, 2003), 287–333.
8. The seemingly largest exception from this rule, the filling of the very large area of South Boston flats, was in fact meant as a harbor improvement, intended to increase the tidal scour's force. Michael Rawson has written extensively about the issue of nineteenth-century attempts to understand and control Boston Harbor: "What Lies Beneath: Science, Nature, and the Making of Boston Harbor," *Journal of Urban History* 35, no. 5 (2009); "What Lies Beneath: Science, Nature, and the Making of Boston Harbor," in *Remaking Boston: An Environmental History of the City and Its Surroundings*, ed. Anthony N. Penna and Conrad E. Wright (Pittsburgh: University of Pittsburgh Press, 2009); *Eden on the Charles: The Making of Boston* (Cambridge, MA: Harvard University Press, 2014), 179–232; and "Rethinking Conservation: The Nineteenth-Century Battle to Save Boston Harbor," in *A Landscape History of New England*, ed. Blake A. Harrison and Richard W. Judd (Cambridge, MA: MIT Press, 2011).
9. For an in-depth account of how the streetcar influenced the nature and pattern of Boston's growth, see the classic study by Sam Bass Warner, *Streetcar Suburbs: The Process of Growth in Boston, 1870–1900* (Cambridge, MA: MIT Press, 1962).
10. Warner, *Streetcar Suburbs*, 4.
11. "Global Boston: A Portal to the Region's Immigrant Past and Present," Boston

College, Department of History, https://globalboston.bc.edu/. For general histories of nineteenth-century immigration to Boston, see, for instance, Oscar Handlin, *Boston's Immigrants, 1790–1880: A Study in Acculturation*, rev. ed. (Cambridge, MA: Belknap Press of Harvard University Press, 1991); Anthony Bak Buccitelli, *City of Neighborhoods: Memory, Folklore, and Ethnic Place in Boston* (Madison: University of Wisconsin Press, 2016); Thomas H. O'Connor, *The Boston Irish: A Political History* (Boston: Northeastern University Press, 1995); and Jonathan D. Sarna and Ellen Smith, eds., *The Jews of Boston: Essays on the Occasion of the Centenary (1895–1995) of the Combined Jewish Philanthropies of Greater Boston* (Boston: Philanthropies, distributed by Northeastern University Press, 1995).

12. Annie Haven Thwing, *The Crooked and Narrow Streets of the Town of Boston, 1630–1822* (Boston: Marshall Jones, 1920); John P. Marquand, *The Late George Apley: A Novel in the Form of a Memoir* (Boston: Little, Brown, 1937), 26. The scene quoted here has often served as an illustration of the social stratification of Boston's geography. See, for instance, Walter Muir Whitehill and Lawrence W. Kennedy, *Boston: A Topographical History*, 3rd ed. (Cambridge, MA: Belknap Press of Harvard University Press, 2000), 119–20; or Theodore G. Clarke, *Beacon Hill, Back Bay, and the Building of Boston's Golden Age* (Charleston, SC: History Press, 2010).
13. Board of Health of the City of Boston, *Tenth Annual Report: City Document No. 87. Documents of the City of Boston for the Year 1882* (Boston: Rockwell and Churchill, City Printers, 1883), 65–66.
14. Thomas H. O'Connor, *The Hub: Boston Past and Present* (Boston: Northeastern University Press, 2001), 94–96; Seasholes, *Gaining Ground*, 292.
15. Lawrence W. Kennedy, *Planning the City upon a Hill: Boston since 1630* (Amherst: University of Massachusetts Press, 1992), 48.
16. "City Point's Claims," *Boston Daily Globe*, August 11, 1873.
17. William B. Meyer, "A City (Only Partly) on a Hill: Terrain and Land Use in Pre-Twentieth-Century Boston," in *Remaking Boston*, ed. Penna and Wright, 146.
18. Jacqueline Barbara Carr, *After the Siege: A Social History of Boston, 1775–1800* (Boston: Northeastern University Press, 2005), 161; Kennedy, *Planning the City upon a Hill*, 44.
19. For more information on King Philip's War and the internment of Native Americans on Deer Island, see Jill Lepore, *The Name of War: King Philip's War and the Origins of American Identity* (New York: Vintage Books, 1999); James D. Drake, *King Philip's War: Civil War in New England, 1675–1676* (Amherst: University of Massachusetts Press, 1999); Lisa Brooks, *Our Beloved Kin: A New History of King Philip's War* (New Haven, CT: Yale University Press, 2018); and Christine M. DeLucia, *Memory Lands: King Philip's War and the Place of Violence in the Northeast* (New Haven, CT: Yale University Press, 2018).
20. John Ballard Blake, *Public Health in the Town of Boston, 1630–1822* (Cambridge, MA: Harvard University Press, 1959), 18–35.
21. Stephen Coss, *The Fever of 1721: The Epidemic That Revolutionized Medicine and American Politics* (New York: Simon & Schuster, 2016), 62–63.
22. Carl Bridenbaugh, *Cities in the Wilderness: The First Century of Urban Life in America, 1625–1742* (New York: Ronald Press, 1938), 401.
23. Fred B. Rogers, "Dr. Samuel Gelston (1727–82), Variolator, and His Son, Dr. Roland Gelston (1761–1829), Vaccinator, from Nantucket," *Journal of the History of Medicine and Allied Sciences* 27, no. 1 (1972), 82; Blake, *Public Health*, 96.
24. Sharon Seitz and Stuart Miller, *The Other Islands of New York City: A History and Guide*, 3rd ed. (Woodstock, VT: Countryman Press, 2011), 87–88.

25. M. F. Sweetser, *King's Handbook of Boston Harbor*, 3rd ed. (Boston: Moses King, 1889), 221; David Kales, *Boston Harbor Islands: A History of an Urban Wilderness* (Charleston, SC: History Press, 2007), 90.
26. Sweetser, *King's Handbook of Boston Harbor*, 186, 204; "Location of State Prison," *Boston Daily Globe*, December 3, 1905.
27. Chris Burrell, "Long Island Bridge Closed; Homeless Forced to Leave Shelter," *Patriot Ledger*, October 8, 2014. For a discussion of path dependency of municipal infrastructure, see Melosi, *Sanitary City*.
28. Steven M. Rudnick, "Remaking Boston Harbor: Cleaning Up after Ourselves," in *Remaking Boston*, ed. Penna and Wright, 60.
29. Board of Health of the City of Boston, *Ninth Annual Report: City Document No. 87. Documents of the City of Boston for the Year 1881* (Boston: Rockwell and Churchill, City Printers, 1882), 54. One load equaled about 43 cubic feet of waste. The 235,000 loads of rubbish collected in 1881 thus equaled approximately 10 million cubic feet, or 283,000 cubic meters.
30. The out-of-sight, out-of-mind logic seemed to guide most of nineteenth- and twentieth-century urban waste management. See Martin V. Melosi, *Garbage in the Cities: Refuse, Reform, and the Environment*, rev. ed. (Pittsburgh: University of Pittsburgh Press, 2005), 31–35; Roman Köster, *Hausmüll: Abfall und Gesellschaft in Westdeutschland 1945–1990* (Göttingen: Vandenhoeck & Ruprecht, 2017), 13; Susan Strasser, *Waste and Want: A Social History of Trash* (New York: Metropolitan Books, 1999), 271–72; and Simone M. Müller, "The 'Flying Dutchmen': Ships' Tales of Toxic Waste in a Globalized World," in *Out of Sight, Out of Mind: The Politics and Culture of Waste*, ed. Christof Mauch, RCC Perspectives: Transformations in Environment and Society (Munich: Rachel Carson Center, 2016).
31. The use—or misuse—of the ocean as the ultimate sink for their waste by coastal cities seems to be universal. For examples, see Sarah S. Elkind, *Bay Cities and Water Politics: The Battle for Resources in Boston and Oakland* (Lawrence: University Press of Kansas, 1998); Matthew Morse Booker, *Down by the Bay: San Francisco's History between the Tides* (Berkeley: University of California Press, 2013); Melosi, *Garbage in the Cities*; and Craig E. Colten, *An Unnatural Metropolis: Wresting New Orleans from Nature* (Baton Rouge: Louisiana State University Press, 2006).
32. Albert B. Hart, ed., *Hamilton's Itinerarium: Being a Narrative of a Journey from Annapolis, Maryland, through Delaware, Pennsylvania, New York, New Jersey, Connecticut, Rhode Island, Massachusetts and New Hampshire from May to September 1744, by Doctor Alexander Hamilton* (St. Louis: William K. Bixby, 1907), 135.
33. O'Connor, *Hub*, 92; Seasholes, *Gaining Ground*, 74–92.
34. Other North American cities built on the coast and in marshy areas wrestled with similar problems in the nineteenth-century period of explosive growth. For examples, see Booker, *Down by the Bay*, 51–52; Matthew W. Klingle, *Emerald City: An Environmental History of Seattle* (New Haven, CT: Yale University Press, 2007), 67–75; and William Cronon, *Nature's Metropolis: Chicago and the Great West* (New York: W. W. Norton, 1991), 55–58. For an extremely critical view of converting wetlands into urban spaces and on land making from the perspective of destruction of marshlands, see Rodney James Giblett, *Cities and Wetlands: The Return of the Repressed in Nature and Culture* (London: Bloomsbury, 2016).
35. Seasholes, *Gaining Ground*, 33, 78.
36. Whitehill and Kennedy, *Boston*, 145–51; Seasholes, *Gaining Ground*, 175–98.
37. Seasholes, *Gaining Ground*, 16.
38. City of Boston Public Works Department, Contract and Specifications for the Disposal

of Garbage and Refuse for Ten Years from July 1, 1922, Coleman Brothers Incorporated, 1922, folder 143: Coleman Disposal Company, box 2, Law Department Records, City of Boston Archives.

39. Board of Health of the City of Boston, *Ninth Annual Report*, 55.
40. Board of Health of the City of Boston, *Twelfth Annual Report*, 46–47.
41. Board of Health of the City of Boston, *Thirteenth Annual Report: City Document No. 89. Documents of the City of Boston for the Year 1885* (Boston: Rockwell and Churchill, City Printers, 1886), 46.
42. Board of Health of the City of Boston, *Fourteenth Annual Report: City Document No. 31. Documents of the City of Boston for the Year 1886* (Boston: Rockwell and Churchill, City Printers, 1887), 50–51; Robin Nagle, *Picking Up: On the Streets and behind the Trucks with the Sanitation Workers of New York City* (New York: Farrar, Straus and Giroux, 2013), 114.
43. Andrew Donovan, "A Brief History of Sewers in Boston: Histories of the Calf Pasture Pumping Station," University Archives and Special Collections in the Healey Library at UMass Boston, http://blogs.umb.edu/pumpingstation/2013/03/19/a-history-of-sewers-in-boston/.
44. Board of Health of the City of Boston, *Twelfth Annual Report*, 30.
45. Andrew Donovan, "Inside the Calf Pasture Pumping Station, 1888: Histories of the Calf Pasture Pumping Station," http://blogs.umb.edu/pumpingstation/2013/04/02/touring-the-interior-of-the-pumping-station/.
46. The introduction of the metropolitan sewerage system played a major role in the development of political regionalism in Boston's metropolitan area. See Sarah S. Elkind, "Building a Better Jungle: Anti-Urban Sentiment, Public Works, and Political Reform in American Cities, 1880–1930," *Journal of Urban History* 24, no. 1 (1997).
47. Board of Health of the City of Boston, *Twelfth Annual Report*, 57–58.
48. Board of Health of the City of Boston, *Ninth Annual Report*, 54.
49. Health Department of the City of Boston, *Annual Report for the Year 1892: City Document No. 13. Documents of the City of Boston for the Year 1893* (Boston: Rockwell and Churchill, City Printers, 1894), 102.
50. This number is remarkable, but not untypical. For a detailed discussion of cattle in nineteenth-century American cities, see Frederick L. Brown, *The City Is More than Human: An Animal History of Seattle* (Seattle: University of Washington Press, 2016), 69–102.
51. Clay McShane and Joel A. Tarr, *The Horse in the City: Living Machines in the Nineteenth Century* (Baltimore: Johns Hopkins University Press, 2007), 57–70.
52. For instance, the Street Department owned in 1891 290 horses, the Board of Health was in 1883 owner of 211 horses, and the Fire Department used 273 horses in 1891, of which it owned 218. See Street Department of the City of Boston, *Annual Report for the Year 1891* (Boston: Rockwell and Churchill, City Printers, 1892), 313; Board of Health of the City of Boston, *Eleventh Annual Report: City Document No. 107. Documents of the City of Boston for the Year 1883* (Boston: Rockwell and Churchill, City Printers, 1884), 90; and Fire Department of the City of Boston, *Nineteenth Annual Report from December 30, 1890, to January 31, 1891* (Boston: Rockwell and Churchill, City Printers, 1892), v.
53. Clay McShane, "Gelded Age Boston," *New England Quarterly* 74, no. 2 (2001): 276.
54. "Money in Dead Horses," *Boston Daily Globe*, September 14, 1890; McShane, "Gelded Age Boston," 300.
55. Sweetser, *King's Handbook of Boston Harbor*, 174–76.
56. "Big Whale May Not Be Towed to Boston," *Boston Daily Globe*, September 21, 1903.

57. William A. Stokinger, "Spectacle Island Chronology Draft" (unpublished manuscript, May 2008), Massachusetts Department of Conservation and Recreation, Division of Urban Parks and Recreation, Bureau of Ranger Services, 23; "Business Troubles," *Boston Daily Globe*, March 3, 1891.
58. Health Department of the City of Boston, *Annual Report for the Year 1898–99: City Document No. 14. Documents of the City of Boston for the Year 1899* (Boston: Municipal Printing Office, 1900), 40.
59. "Wells under Fire," *Boston Daily Globe*, October 31, 1899; Health Department of the City of Boston, *Annual Report for the Year 1899–1900*, 44–45; "Must Get Out: Mayor Collins after Garbage Plant," *Boston Daily Globe*, August 9, 1903.
60. Health Department of the City of Boston, *Annual Report for the Year 1910–1911: City Document No. 16. Documents of the City of Boston for the Year 1911* (Boston: Printing Department, 1912), 8. Around the same time, Barren Island off Brooklyn was being used for much the same purposes of garbage processing and animal rendering. See Miriam Sicherman, *Brooklyn's Barren Island: A Forgotten History* (Charleston, SC: History Press, 2019).
61. Health Department of the City of Boston, *Annual Report for the Year 1892*, 62–63.
62. Board of Health of the City of Boston, *Sixth Annual Report*, 3.
63. This tendency was not particular to Boston or, indeed, the United States. For European examples, see Anna Mazanik, "Sanitation, Urban Environment and the Politics of Public Health in Late Imperial Moscow" (PhD diss., Central European University, 2015).
64. Board of Health of the City of Boston, *Twelfth Annual Report*, 46, 57; "Garbage Cremation Plant," *Boston Daily Globe*, December 9, 1897.
65. Board of Health of the City of Boston, *Sixth Annual Report*, 40; Board of Health of the City of Boston, *Twelfth Annual Report*, 46–47.
66. Board of Health of the City of Boston, *Sixth Annual Report*, 40.
67. Special Commission Relative to the Discharge of Sewage into Boston Harbor, *Report under Chapter 29* (Boston: Wright and Potter, 1930), Senate No. 56, 51–52.
68. Waste as "matter out of place" is a concept famously coined by the anthropologist Mary Douglas in *Purity and Danger: An Analysis of Concepts of Pollution and Taboo* (London: Routledge & K. Paul, 1966).
69. Germany's leading waste historian, Roman Köster, calls this effect the "Freudian interpretation of consumerism," likening the refuse to suppressed emotions that nevertheless find their way back to the conscious mind to haunt their bearer. See Köster, *Hausmüll*, 13.
70. Board of Health of the City of Boston, *Twelfth Annual Report*, 59–60.
71. Board of Health of the City of Boston, *Fifteenth Annual Report: City Document No. 13. Documents of the City of Boston for the Year 1887* (Boston: Rockwell and Churchill, City Printers, 1888), 42.
72. "Controversy as to City Garbage," *Boston Daily Globe*, May 10, 1905.
73. "More Scows Needed," *Boston Daily Globe*, February 2, 1896.
74. "New Channel as Dumping Ground," *Boston Daily Globe*, June 12, 1903.
75. Joint Board Acting under the Provisions of Chapter 56 of the Resolves of 1917, *Report on an Investigation of the Discharge of Sewage into Boston Harbor* (Boston: Wright and Potter, 1918), House No. 1215; Special Commission Relative to the Discharge of Sewage into Boston Harbor, *Report under Chapter 29*; Special Commission on the Discharge of Sewage into Boston Harbor and Its Tributaries, *Report, December 1936* (Boston: Wright and Potter, 1937), House No. 1600.
76. Joint Board Acting under the Provisions of Chapter 56 of the Resolves of 1917, *Report on an Investigation of the Discharge of Sewage into Boston Harbor*, 5.

77. Special Commission Relative to the Discharge of Sewage into Boston Harbor, *Report under Chapter 29*, 53.
78. On the topic of the olfactory aspect of urban horses, see Brown, *City Is More than Human*, 121–22; and McShane and Tarr, *Horse in the City*, 26–27.
79. In the late nineteenth and early twentieth centuries, Spectacle Island was specifically mentioned as a nuisance in the following annual reports of the Boston Board of Health: Board of Health of the City of Boston, *Second Annual Report*, 13–20; Board of Health of the City of Boston, *Third Annual Report: City Document No. 85. Documents of the City of Boston for the Year 1875* (Boston: Rockwell and Churchill, City Printers, 1876), 17–18; Health Department of the City of Boston, *Annual Report for the Year 1901–1902: City Document No. 17. Documents of the City of Boston for the Year 1902* (Boston: Municipal Printing Office, 1903), 47; Health Department of the City of Boston, *Annual Report for the Year 1906–1907: City Document No. 19. Documents of the City of Boston for the Year 1907* (Boston: Municipal Printing Office, 1908), 49; Health Department of the City of Boston, *Annual Report for the Year 1908–1909*, 63–64; Health Department of the City of Boston, *Annual Report for the Year 1910–1911*, 8; Health Department of the City of Boston, *Annual Report for the Year 1921: City Document No. 12. Documents of the City of Boston for the Year 1922* (Boston: Printing Department, 1923), 113; Health Department of the City of Boston, *Annual Report for the Year 1923: City Document No. 12. Documents of the City of Boston for the Year 1924* (Boston: Printing Department, 1925), 103; Health Department of the City of Boston, *Annual Report for the Year 1924: City Document No. 13. Documents of the City of Boston for the Year 1925* (Boston: Printing Department, 1926), 134; and Health Department of the City of Boston, *Fifty-Sixth Annual Report for the Year 1927: City Document No. 12. Documents of the City of Boston for the Year 1928* (Boston: Printing Department, 1929), 8.
80. Kiechle, *Smell Detectives*, 5.
81. Board of Health of the City of Boston, *Second Annual Report*, 18.
82. "Look to Council," *Boston Daily Globe*, October 10, 1901.
83. "Bad Day for South Boston," *Boston Daily Globe*, March 29, 1907; "Complaint of Garbage Plant," *Boston Daily Globe*, September 29, 1911; "Editorial Points," *Boston Daily Globe*, July 24, 1913.
84. Health Department of the City of Boston, *Annual Report for the Year 1925: City Document No. 13. Documents of the City of Boston for the Year 1926* (Boston: Printing Department, 1927), 131.
85. "Complain of Foul Odors," *Boston Daily Globe*, September 16, 1907.
86. "Wind Brings Strong Odor and Wakens Quincy Folk," *Boston Daily Globe*, August 11, 1928.
87. "South Boston," Global Boston, https://globalboston.bc.edu.
88. Health Department of the City of Boston, *Annual Report for the Year 1901–1902*, 47.
89. "Winthrop Threatens Boston with Court Action If Necessary," *Boston Daily Globe*, July 20, 1922.
90. Robert H. Wyatt, "A Short Description of Spectacle Island and What Transpired There as Remembered by Me" (unpublished manuscript, 2001), Suzanne Gall Marsh private collection.
91. "Must Get Out."
92. "Better Water," *Boston Daily Globe*, May 9, 1895.
93. Board of Health of the City of Boston, *Fourth Annual Report*, 44.
94. Seasholes, *Gaining Ground*, 338.
95. Sweetser, *King's Handbook of Boston Harbor*, 108.
96. Donald G. Jones, *Intensive Archaeological Survey of Spectacle Island in Boston Harbor*,

with the assistance of Nancy S. Seasholes (Boston: Office of Public Archaeology, Boston University, 1989), 12; Sweetser, *King's Handbook of Boston Harbor*, 174.

97. Nathaniel B. Shurtleff, *A Topographical and Historical Description of Boston* (Boston: City Council, 1871), 517.
98. "Boston's Only District School," *Boston Daily Globe*, December 1, 1901.
99. Louis E. Cellineri, *Seaport Dynamics: A Regional Perspective* (Lexington, MA: Lexington Books, 1976), 16.
100. City of Boston Public Works Department, Contract and Specifications for the Disposal of Garbage and Refuse for Ten Years from July 1, 1922, Law Department Records, City of Boston Archives.
101. Jones, *Intensive Archaeological Survey of Spectacle Island*, 17.
102. George C. Hyland, Commissioner of Public Works, to John Tobin, Corporation Counsel, to Boston Law Department, October 30, 1941, folder Spectacle Island, Boston Harbor, box 45, Law Department Records, City of Boston Archives.
103. Jones, *Intensive Archaeological Survey of Spectacle Island*, 2, 17.
104. Contract for the Disposal of Garbage and Refuse for the Year 1948, Coleman Disposal Company, folder Coleman Disposal, box 41, Law Department Records, City of Boston Archives.
105. Shurtleff, *Topographical and Historical Description of Boston*, 517.
106. "Island Schoolhouse," *Boston Daily Globe*, October 23, 1898.
107. David E. Nye, "The Anti-Landscape," in *The Anti-Landscape*, ed. David E. Nye and Sarah S. Elkind (Amsterdam: Rodopi, 2014), 11–26.
108. Board of Health of the City of Boston, *Tenth Annual Report*, 66.
109. Board of Health of the City of Boston, *Fourth Annual Report*, 44; Board of Health of the City of Boston, *Fifth Annual Report: City Document No. 67. Documents of the City of Boston for the Year 1877* (Boston: Rockwell and Churchill, City Printers, 1878), 64.
110. Jones, *Intensive Archaeological Survey of Spectacle Island*, 13; "Island Schoolhouse"; "Boston's Only District School."
111. Stokinger, "Spectacle Island Chronology Draft," 25, 43.
112. Wyatt, "Short Description of Spectacle Island"; Kathleen A. Rivera, interview by the author, October 21, 2016, Quincy, MA.
113. Rivera, interview by the author; Jenny Lescovitz quoted by Kathleen A. Rivera, interview by Boston Harbor Islands Mass. Memories Road Show, September 17, 2011, University Archives and Special Collections, Joseph P. Healey Library, University of Massachusetts, Boston.
114. Wyatt, "Short Description of Spectacle Island."
115. Billie Wyatt-Hargreaves, diary, transcribed by Jeff Peters, June 2004, Spectacle Island Interpretive Guide, City of Boston Archives.
116. "Harbor Police Find 153 Live on Islands," *Boston Globe*, January 3, 1937.
117. George G. Hyland, Commissioner of Public Works, to John Tobin, Corporation Counsel to Boston Law Department, October 30, 1941; John Tobin to Maurice J. Tobin, Mayor of Boston, November 4, 1941; City of Boston Board of Assessors: 1941 Assessment Record of Spectacle Island, November 8, 1941, all in folder Spectacle Island, Boston Harbor, box 45, Law Department Records, City of Boston Archives.
118. For examples of such recommendations, see Health Department of the City of Boston, *Annual Report for the Year 1903–1904: City Document No. 17. Documents of the City of Boston for the Year 1904* (Boston: Municipal Printing Office, 1905), 49; Health Department of the City of Boston, *Annual Report for the Year 1905–1906*, 41; and Public Works Department of the City of Boston, *Annual Report for the Year 1938* (Boston: Printing Department, 1939), 11.

119. Finance Commission of the City of Boston to Maurice J. Tobin, Mayor of the City of Boston, August 13, 1941, folder Spectacle Island, Boston Harbor, box 45, Law Department Records, City of Boston Archives.
120. "Being Modern," *Boston Globe*, August 18, 1944.
121. Public Works Department of the City of Boston, *Annual Report for the Year 1959* (Boston: Printing Department, 1960), 49.

Chapter 3: The City of Tomorrow

1. Derek Lamb and Lawrence Rosenblum, *Boston: Harbor/City/Islands*, produced by Stanley Jacks, available courtesy of University Archives and Special Collections, Joseph P. Healey Library, University of Massachusetts, Boston, SC-0175: Volunteers and Friends of the Boston Harbor Islands Inc. records (Cambridge, MA: Massachusetts Institute of Technology, 1969); Lawrence Rosenblum, interview by the author, December 14, 2020.
2. William B. Meyer, "A City (Only Partly) on a Hill: Terrain and Land Use in Pre-Twentieth-Century Boston," in *Remaking Boston: An Environmental History of the City and Its Surroundings*, ed. Anthony N. Penna and Conrad E. Wright (Pittsburgh: University of Pittsburgh Press, 2009), 146.
3. Aubrey Parkman, *Army Engineers in New England: The Military and Civil Work of the Corps of Engineers in New England, 1775–1975* (Waltham, MA: U.S. Army Corps of Engineers, 1978), 10–25, 107–32; Gerald Butler, *The Military History of Boston's Harbor Islands* (Charleston, SC: Arcadia, 2000).
4. Nancy S. Seasholes, *Gaining Ground: A History of Landmaking in Boston* (Cambridge, MA: MIT Press, 2003), 295.
5. Louis E. Cellineri, *Seaport Dynamics: A Regional Perspective* (Lexington, MA: Lexington Books, 1976), 23, 26.
6. Lawrence W. Kennedy, *Planning the City upon a Hill: Boston since 1630* (Amherst: University of Massachusetts Press, 1992), 110.
7. Cellineri, *Seaport Dynamics*, 18.
8. John R. Gillis, *The Human Shore: Seacoasts in History* (Chicago: University of Chicago Press, 2012), 113.
9. Cellineri, *Seaport Dynamics*, 93–94.
10. For wider cultural history implications of this shift, see Steven Mentz, "Toward a Blue Cultural Studies: The Sea, Maritime Culture, and Early Modern English Literature," *Literature Compass* 6, no. 5 (2009): 228. For an economic history of containerization, see Marc Levinson, *The Box: How the Shipping Container Made the World Smaller and the World Economy Bigger*, 2nd ed. (Princeton, NJ: Princeton University Press, 2016).
11. For a detailed history of Logan Airport expansion, see Seasholes, *Gaining Ground*, 375–83.
12. For accounts of Boston shipping and its decline in the early twentieth century, see Cellineri, *Seaport Dynamics*; and Works Project Administration, *Boston Looks Seaward: The Story of the Port, 1630–1940* (Boston: B. Humphries, 1941).
13. Thomas H. O'Connor, *Building a New Boston: Politics and Urban Renewal, 1950–1970* (Boston: Northeastern University Press, 1993), 83.
14. Thomas H. O'Connor, *The Hub: Boston Past and Present* (Boston: Northeastern University Press, 2001), 207.
15. Kennedy, *Planning the City upon a Hill*, 157.

16. Matthew Edel, Elliott Sclar, and Dan Luria, *Shaky Palaces: Homeownership and Social Mobility in Boston's Surburbanization* (New York: Columbia University Press, 1984), 74, 67.
17. O'Connor, *Building a New Boston*, 67.
18. O'Connor, *Hub*, 209. On Boston's suburbanization, see James C. O'Connell, *The Hub's Metropolis: Greater Boston's Development from Railroad Suburbs to Smart Growth* (Cambridge, MA: MIT Press, 2013).
19. Metropolitan Area Planning Council, *Issues and Problems of Boston Metropolitan Area Development* (Cambridge, MA: Joint Center for Urban Studies of the Massachusetts Institute of Technology and Harvard University, 1965), 11–12.
20. O'Connor, *Building a New Boston*, 147.
21. O'Connor, *Hub*, 208; Kennedy, *Planning the City upon a Hill*, 151; O'Connor, *Building a New Boston*, 15.
22. Samuel Zipp and Michael Carriere, "Introduction: Thinking through Urban Renewal," *Journal of Urban History* 39, no. 3 (2012): 359.
23. O'Connor, *Building a New Boston*, 42–46.
24. O'Connor, *Hub*, 216; O'Connor, *Building a New Boston*, 56.
25. Kennedy, *Planning the City upon a Hill*, 161; O'Connor, *Building a New Boston*, 190.
26. Boston Redevelopment Authority, *1965/1975 General Plan for the City of Boston and the Regional Core* (Boston, 1965), 1.
27. It goes beyond the scope of this book to offer a comprehensive discussion of the various political, social, environmental, and intellectual aspects of urban renewal in the postwar United States. There exists a vast literature about this topic that has experienced a new surge of interest in recent years. Relevant recent studies include Alison Isenberg, *Designing San Francisco: Art, Land, and Urban Renewal in the City by the Bay* (Princeton, NJ: Princeton University Press, 2017); Samuel Zipp, *Manhattan Projects: The Rise and Fall of Urban Renewal in Cold War New York* (Oxford: Oxford University Press, 2010); Donald Craig Parson, *Making a Better World: Public Housing, the Red Scare, and the Direction of Modern Los Angeles* (Minneapolis: University of Minnesota Press, 2005); Lydia R. Otero, *La Calle: Spatial Conflicts and Urban Renewal in a Southwest City* (Tucson: University of Arizona Press, 2010); and Robert O. Self, *American Babylon: Race and the Struggle for Postwar Oakland* (Princeton, NJ: Princeton University Press, 2005). A recent study offering a reappraisal of the conventional narrative of urban renewal as an unmitigated disaster is Lizabeth Cohen, *Saving America's Cities: Ed Logue and the Struggle to Renew Urban America in the Suburban Age* (New York: Farrar, Straus and Giroux, 2019).
28. O'Connor, *Hub*, 223.
29. O'Connor, *Building a New Boston*, 124, 127. The most detailed study of the political, social, and cultural history of the West End and its demolition is Sean M. Fisher and Carolyn Hughes, eds., *The Last Tenement: Confronting Community and Urban Renewal in Boston's West End* (Boston: Bostonian Society, 1992).
30. Boston Harbor Islands Commission Conference, 1968–69, Conference Transcript—Workshop 5: Community Development, June 24, 1969, folder 13, box 2, series II, David C. Harrison Papers, State Library of Massachusetts Special Collections.
31. O'Connor, *Building a New Boston*, 215.
32. Michael Rawson, "What Lies Beneath: Science, Nature, and the Making of Boston Harbor," *Journal of Urban History* 35, no. 5 (2009): 679, 692.
33. "Old Boston Would Not Know Herself If All This Happened," *Boston Daily Globe*, February 11, 1907.
34. "Second Interim Report of the Special Commission on the Boston Harbor Islands: Massachusetts House of Representatives Resolve No. 4884" (Boston, 1970), 11.

35. "Legislature to Study Harbor Islands Use," *Boston Globe*, September 2, 1966; "World's Fair Seen as Key: Expansion into Harbor Could Add 150,000 Islanders," *Boston Globe*, November 13, 1966.
36. *Interim Report of the Special Commission on the Boston Harbor Islands: Massachusetts House of Representatives Resolve No. 5457* (Boston: Wright and Potter, 1969), 6.
37. Seasholes, *Gaining Ground*.
38. "Second Interim Report of the Special Commission on the Boston Harbor Islands," 21.
39. For examples, see Boston Redevelopment Authority, *1965/1975 General Plan for the City of Boston and the Regional Core*, 7; *Interim Report of the Special Commission on the Boston Harbor Islands*; *Third and Final Report of the Special Commission on the Boston Harbor Islands: Massachusetts House of Representatives Resolve No. 5051* (Boston: Wright and Potter, 1970); Edward J. Logue, *Seven Years of Progress: A Final Report* (Boston: Boston Redevelopment Authority, 1967), 30; Fred Pillsbury, "White Proposes Program to Develop Harbor, Islands," *Boston Globe*, April 21, 1970; and Robert E. Levey, "Boston's Neglected Jewels," *Boston Globe*, April 18, 1967.
40. James B. Ayres, "MDC Urged to Turn Peddock's Island into Recreation Area," *Boston Globe*, August 13, 1967; "Peddock's Island—a Sad Waste," *Boston Globe*, September 8, 1968.
41. Pillsbury, "White Proposes Program to Develop Harbor, Islands"; Levey, "Boston's Neglected Jewels."
42. *Interim Report of the Special Commission on the Boston Harbor Islands*, 7.
43. The other eight were the Government Center, Washington Park, Charlestown, South End, South Cove, Fenway, Central Business District, and Madison Park. See Logue, *Seven Years of Progress*, 2.
44. Logue, *Seven Years of Progress*, 30–32.
45. Metropolitan Area Planning Council, *Open Space and Recreation Plan and Program for Metropolitan Boston*, vol. 1 (Boston: MAPC, 1969), 9, 16, 37.
46. Boston Redevelopment Authority, *1965/1975 General Plan for the City of Boston and the Regional Core*, 75.
47. "Newport's Goat Island Development Sets an Example for Boston," *Boston Globe*, July 26, 1970.
48. Joseph A. Keblinski, "Dumping Problem Dumped on Hub," *Boston Globe*, July 8, 1962.
49. Fred Pillsbury, "Owners All Favor Recreation: Private Islands Would Adapt to Any Plan," *Boston Globe*, November 24, 1968.
50. Boston Harbor Islands Commission Conference, 1968–69, Conference Transcript—Workshop 2: Transportation, June 24, 1969, folder 10, box 1, series II, Harrison Papers, State Library of Massachusetts Special Collections.
51. Meeting of Special Legislative Commission Charged with the Duty of Investigating and Studying the Possibility of Developing the Islands of Boston Harbor, November 1, 1967, State Library of Massachusetts Special Collections, 25.
52. Boston Harbor Islands Commission Conference, 1968–69, Conference Transcript—Workshop 2: Transportation.
53. Meeting of Special Legislative Commission Charged with the Duty of Investigating and Studying the Possibility of Developing the Islands of Boston Harbor, November 1, 1967, 36.
54. Seasholes, *Gaining Ground*, 375–82.
55. William P. Coughlin, "Beware of Touching Hub's Outer Islands," *Boston Globe*, January 24, 1971.
56. Boston Harbor Islands Commission Conference, 1968–69, Conference Transcript—Workshop 2: Transportation.

57. The specific 1960s brand of technological optimism applied to exploration and development of ocean space is explored in Helen M. Rozwadowski, "Engineering, Imagination, and Industry: Scripps Island and Dreams for Ocean Science in the 1960s," in *The Machine in Neptune's Garden: Historical Perspectives on Technology and the Marine Environment*, ed. Helen M. Rozwadowski and David K. van Keuren (Sagamore Beach, MA: Science History, 2004).
58. Maurice J. Ruby, "What People Talk About: There's Tax Gold to Be Had by Developing Harbor Islands," *Boston Globe*, April 20, 1960.
59. James B. Ayres, "Time for a Change: Boston's Harbor—a Problems Dump," *Boston Globe*, February 20, 1966; "Island City Near Boston Proposed," *Christian Science Monitor*, July 29, 1967.
60. "World's Fair Seen as Key."
61. "Linked Harbor Islands Could Provide Fairground: N.E. Governors Like Hub as '75 Fair Site," *Boston Globe*, October 17, 1966.
62. Daniel J. Ahern, Martin A. Adler, and David Crane to Edward J. Logue, November 15, 1963, folder 23: 1975 World Freedom Fair, Inc., Memoranda 1963, series III, Harrison Papers, State Library of Massachusetts Special Collections.
63. World Fair Project—Draft of Boston Redevelopment Authority Annual Report, February 24, 1967, folder 27, series III, Harrison Papers, State Library of Massachusetts Special Collections.
64. Anthony J. Yudis, "Several Sites Eyed for World Fair Here," *Boston Globe*, November 16, 1964.
65. "Boston's Freedom 75 May Spawn a City," *Boston Traveler*, June 13, 1967.
66. Ahern, Adler, and Crane to Logue, November 15, 1963.
67. Plan for the United States Bicentennial World Exhibition Boston, Massachusetts, April–October 1975, 1975 Freedom Fair, Inc./Boston Redevelopment Authority, David C. Harrison, September 30, 1966, Harbor Islands N622 P, Boston Public Library.
68. "Legislature to Study Harbor Islands Use"; "Linked Harbor Islands Could Provide Fairground."
69. The cultural imagination of islands as utopias is discussed in John R. Gillis, *Islands of the Mind: How the Human Imagination Created the Atlantic World* (New York: Palgrave Macmillan, 2004), 75–76; Uwe Lübken and Rebecca Hofmann, "Laboratorien der ökologischen Moderne? Umwelt, Wissen und Geschichte (auf) der kleinen Insel," *Aus Politik und Zeitgeschichte* 68, nos. 32–33 (2018): 4; Arndt Kremer, "Eine Welt für sich: Die Insel als literarischer und sprachlicher Grenz- und Denkraum," *Aus Politik und Zeitgeschichte* 68, nos. 32–33 (2018): 47–52.
70. John Opie, *Virtual America: Sleepwalking through Paradise* (Lincoln: University of Nebraska Press, 2008), 83.
71. David C. Harrison, Selected Applications of Research and Development and Other Ideas to the U.S. Bicentennial Experimental Prototype Urban Community in Dorchester Bay, July 1967, folder 34: U.S. Bicentennial Experimental Prototype Urban Community in Dorchester Bay, July 1967, series III, Harrison Papers, State Library of Massachusetts Special Collections.
72. Claire Poitras, "A City on the Move: The Surprising Consequences of Highways," in *Metropolitan Natures: Environmental Histories of Montreal*, ed. Stéphane Castonguay and Michèle Dagenais (Pittsburgh: University of Pittsburgh Press, 2011), 171.
73. "Philadelphia and Boston Are at War," *New York Times*, June 22, 1969.
74. Anthony J. Yudis, "54 Attend Expo to Plan Boston Fair," *Boston Globe*, June 9, 1967.
75. David C. Harrison, Notes on Conversations with Expo 67 Personnel, February 23, 1966,

folder 26: 1975 World Freedom Fair, Inc., Notes, February 1966, series III, Harrison Papers, State Library of Massachusetts Special Collections.

76. Ahern, Adler, and Crane to Logue, November 15, 1963; Freedom 75, *The United States Bicentennial Celebration: A World Freedom Fair in Boston, April–October 1975* (Boston, 1966), 14.
77. *United States Bicentennial Celebration*; "Our History," Thompson Island Outward Bound Education Center, https://thompsonisland.org.
78. Plan for the United States Bicentennial World Exhibition Boston, 23, 29, 32–33; Donald White, "It Takes Money to Snare a Fair," *Boston Globe*, December 24, 1967.
79. Plan for the United States Bicentennial World Exhibition Boston, 41.
80. Jan Wampler, personal communications and interviews by the author, November 19, 21, and 27, 2020; Jan Wampler, "Projecting the New Boston: From the MBTA to Expo '76," lecture, Bostonian Society of Architects, December 8, 2018, YouTube, https://www.youtube.com/watch?v=7QfFl41dEFo; Planning Design Group, "Text of Slide Presentation: 1976 International Exposition and New Community," Boston Redevelopment Authority, December 23, 1968, Boston Public Library.
81. The prominent urban planner and theorist Donald Appleyard was an MIT graduate and stayed at his alma mater as assistant professor for six years after graduation. During this time, he cooperated with the MAPC on regional planning for the Boston metropolitan area. See Appleyard, "Metropolitan Boston: An Environment for Human Development," in *Issues and Problems of Boston Metropolitan Area Development*, subsection "Physical Planning for the Boston Metropolitan Area" (Boston: MAPC, 1965), 45–58.
82. Planning Design Group, Urban Design Department, and Boston Redevelopment Authority, "United States Bicentennial World Exposition, Boston 1976: New Community for Boston," preliminary summary report, August 1968; Planning Design Group, Urban Design Department, and Boston Redevelopment Authority, "1976 International Exposition and New Community," preliminary summary report, November 1968; Mark Kubo, Chris Grimley, and Mark Pasnik, "Futures: Expo Boston '76," *MasContext*, no. 18 (Summer 2013), http://www.mascontext.com/issues/18-improbable-summer-13/futures-expo-boston-76/.
83. "1976 International Exposition and New Community," 5–15.
84. Wampler, interview by the author, November 27, 2020; Planning Design Group, "Text of Slide Presentation"; "1976 International Exposition and New Community."
85. "United States Bicentennial World Exposition, Boston 1976"; Ayres, "Time for a Change."
86. Ayres, "Time for a Change."
87. Ecological Studies of Boston Harbor, Miscellaneous Papers and Other Harbor Studies, Harbor Islands B65R, Boston Public Library.
88. Richard Powers, "Southies Are Upset over Plans for Expo," *Boston Globe*, April 13, 1969.
89. Barry Cadigan, "Yachtsmen Sponsor Save-Harbor-Islands Cruise," *Boston Globe*, July 21, 1968; Barry Cadigan, "Cruise Points Out Threat to Boston Harbor," *Boston Globe*, August 3, 1969.
90. Cadigan, "Yachtsmen Sponsor Save-Harbor-Islands Cruise."
91. Robert L. Turner, "Harbor Study Chairman Calls Expo '76 Plan Impractical," *Boston Globe*, March 4, 1969.
92. Ian S. Menzies, "Boston Still Has Stake in Expo '76," *Boston Globe*, July 14, 1970.
93. Kubo, Grimley, and Pasnik, "Futures: Expo Boston '76"; "Expo '76 Protest to Mount Tonight," *Boston Globe*, May 21, 1969; Menzies, "Boston Still Has Stake in Expo '76."

94. "Hub Needs a Jab . . . Like a Fair in '75," *Boston Traveler*, June 15, 1967; Powers, "Southies Are Upset over Plans for Expo."
95. David R. Ellis, "Boston Council Opposes Expo 76," *Boston Globe*, September 5, 1969.
96. "Boston to Drop Drive for '76 Bicentennial," *New York Times*, September 14, 1969.
97. Gerald M. O'Neill, "Hub's Expo 76 Rejected, 8–0, by City Council," *Boston Globe*, September 9, 1969.
98. Robert M. Smith, "Boston, Washington and Philadelphia Put in Bids for Site of American Expo in 1976," *New York Times*, September 25, 1969.
99. Robert J. Sales, "Expo '76 Is Alive—but Hidden Away in a Room at City Hall," *Boston Globe*, January 24, 1970. Philadelphia's expo plan was rejected by the federal government's American Revolution Bicentennial Commission in 1969, and Philadelphia mayor Frank Rizzo canceled the city's international exposition bid in 1972. See Madison Egger Crowe and Scott Gabriel Knowles, "Bicentennial (1976)," in *The Encyclopedia of Greater Philadelphia*, https://philadelphiaencyclopedia.org/archive/bicentennial-1976/.
100. Leonard J. Walsh, "Expo—No," *Boston Globe*, March 31, 1969.
101. "To Save Boston Harbor," *Boston Globe*, August 10, 1969.
102. Sales, "Expo '76 Is Alive."
103. M. F. Sweetser, *King's Handbook of Boston Harbor*, 3rd ed. (Boston: Moses King, 1889), 176.
104. *A Descriptive and Historical Sketch of Boston Harbor and Surroundings* (Boston: W. M. Tenney, 1885), 69, 84.
105. James H. Stark, *Illustrated History of Boston Harbor* (Boston: Photo-Electrotype, 1879), 121, 97; Suzanne Buchanan, "A Crow Point Cottage: Out of the Archives; Exploring the Archives at Hingham Heritage Museum," https://outofthearchives.org/2016/09/27/a-crow-point-cottage/.
106. Stark, *Illustrated History of Boston Harbor*, 98, 123.
107. "The Marine Park at City Point," *Boston Daily Advertiser*, May 28, 1881.
108. For a detailed discussion of park construction as a means of social control in nineteenth-century America, see Paul S. Boyer, *Urban Masses and Moral Order in America, 1820–1920* (Cambridge, MA: Harvard University Press, 1978); David Schuyler, *The New Urban Landscape: The Redefinition of City Form in Nineteenth-Century America* (Baltimore: Johns Hopkins University Press, 1986); and Stephen Hardy, *How Boston Played: Sport, Recreation, and Community, 1865–1915* (Knoxville: University of Tennessee Press, 2003).
109. Seasholes, *Gaining Ground*, 326.
110. "Bits of Paradise: Near Boston and in Danger of Destruction," *Boston Daily Globe*, February 3, 1893; "Revere Beach: Delighted Crowds There Every Day," *Boston Daily Globe*, August 16, 1908.
111. "More than 150,000 Visit Revere Beach during Day," *Boston Globe*, June 23, 1930.
112. Board of Commissioners of the Department of Parks for the City of Boston, *Twenty-Fourth Annual Report for the Year Ending January 31, 1901* (Boston: City of Boston, 1901), 13; Metropolitan Area Planning Council, *Open Space and Recreation Study for the Boston Region* (Boston: Metropolitan Area Planning Council, 1965).
113. A similar tendency to redevelop urban beaches as in Boston could be observed, for instance, in Los Angeles. See Elsa Devienne, "Urban Renewal by the Sea: Reinventing the Beach for the Suburban Age in Postwar Los Angeles," *Journal of Urban History* 45, no. 1 (2019).
114. Metropolitan Area Planning Council, Metropolitan District Commission and Department of Natural Resources, *Open Space and Recreation Plan and Program*

for Metropolitan Boston, vol. 1 (Boston: Commonwealth of Massachusetts, 1969), 5; *Preliminary Report of the Department of Natural Resources Relative to the Further Development of the Public Forests, Parks and Other Lands of the Commonwealth under Chapter 471, Acts of 1955: Massachusetts Senate Resolution No. 740* (Boston: Wright and Potter, 1957), 14; Boston Redevelopment Authority, *1965/1975 General Plan for the City of Boston and the Regional Core*, 74.

115. *Preliminary Report of the Department of Natural Resources*, 138, 18, 19.
116. Metropolitan Area Planning Council, Metropolitan District Commission and Department of Natural Resources, *Open Space and Recreation Plan and Program*, 1:83.
117. Gillis, *Human Shore*, 114–15.
118. James B. Ayres, "Let State Buy Hub Islands?," *Boston Globe*, March 27, 1969.
119. Metropolitan Area Planning Council, Metropolitan District Commission and Department of Natural Resources, *Open Space and Recreation Plan and Program*, 1:9.
120. James B. Ayres, "31 Islands to Play on," *Boston Globe*, March 31, 1967.
121. "Legislature to Study Harbor Islands Use."
122. A. S. Plotkin, "Boston Take Note: Ontario Points Way to Recreational Use of Islands," *Boston Globe*, September 10, 1967.
123. John Harris, "Ring of Fun around Toronto," *Boston Globe*, July 21, 1968.
124. Harbor Islands Study Group, Massachusetts Institute of Technology, *The Boston Harbor Islands* (Cambridge, MA: MIT, 1969), 1, 37–46.
125. "One Island Step," *Boston Globe*, May 14, 1970, 26.
126. The conference proceedings were printed as *Third and Final Report of the Special Commission on the Boston Harbor Islands*.
127. *Third and Final Report*, 28; Boston Harbor Islands Commission Conference, 1968–69, Conference Transcript—Workshop 5: Community Development.
128. Boston Harbor Islands Commission Conference, 1968–69, Conference Transcript—Workshop 4: Ecology/Oceanology, June 24, 1969, folder 12, box 2, series II, Harrison Papers, State Library of Massachusetts Special Collections.
129. James B. Ayres, "Purchase of 16 Boston Islands Backed," *Boston Globe*, August 12, 1969.
130. Boston Harbor Islands Commission Conference, 1968–69, Conference Transcript—Workshop 4: Ecology/Oceanology.
131. Lamb and Rosenblum, *Boston: Harbor/City/Islands*.
132. Dale F. Levering, *An Illustrated Flora of the Boston Harbor Islands* (Boston: Northeastern University Press, 1978), 1, 191–97.
133. Julie A. Richburg and William A. Patterson, "Historical Description of the Vegetation of the Boston Harbor Islands, 1600–2000," *Northeastern Naturalist* 12, no. 3 (2005): 24.
134. Alan Mikal, *Exploring Boston Harbor* (North Quincy, MA: Christopher, 1973), 81.
135. "Opening the Islands," *Boston Globe*, August 4, 1975.

Chapter 4: The Romance of Boston Bay

1. Edward Rowe Snow, *Secrets of the North Atlantic Islands* (New York: Dodd, Mead, 1950), 70. The description of the tour is a compilation from various renditions, chiefly Stephanie Schorow, *East of Boston: Notes from the Harbor Islands* (Charleston, SC: History Press, 2008), 39; and Suzanne Gall Marsh, interview by Boston Harbor Islands Mass. Memories Road Show, September 17, 2011, University Archives and Special Collections, Joseph P. Healey Library, University of Massachusetts, Boston.
2. Edward Rowe Snow told and retold the Lady in Black story in many of his books.

Although the retellings differ from each other in details, the core of the story remains the same. Examples include *Historic Fort Warren* (Boston: Yankee, 1941); *The Romance of Boston Bay* (Boston: Yankee, 1944); *Secrets of the North Atlantic Islands*; and *The Islands of Boston Harbor, 1630–1971*, 2nd ed. (New York: Dodd, Mead, 1971).

3. Schorow, *East of Boston*, 39; Snow, *Secrets of the North Atlantic Islands*, 77.
4. For examples, see Emily Kales and David Kales, *All about the Boston Harbor Islands* (Boston: Herman, 1976), 61; Christopher Klein, *Discovering the Boston Harbor Islands: A Guide to the City's Hidden Shores*, 2nd ed. (Boston: Union Park Press, 2011), 31–32; David Kales, *Boston Harbor Islands: A History of an Urban Wilderness* (Charleston, SC: History Press, 2007), 85–86; and Schorow, *East of Boston*, 40. For examples of Boston Harbor ghost stories, see Holly Mascott Nadler, *Ghosts of Boston Town: Three Centuries of True Hauntings* (Camden, ME: Down East Books, 2002); and Sam Baltrusis, *Haunted Boston Harbor* (Charleston, SC: History Press, 2016).
5. Snow, *Islands of Boston Harbor, 1630–1971*, 35.
6. The story does not appear in Nathaniel B. Shurtleff, *A Topographical and Historical Description of Boston* (Boston: City Council, 1871); M. F. Sweetser, *King's Handbook of Boston Harbor*, 3rd ed. (Boston: Moses King, 1889); James H. Stark, *Illustrated History of Boston Harbor* (Boston: Photo-Electrotype, 1879); or *A Descriptive and Historical Sketch of Boston Harbor and Surroundings* (Boston: W. M. Tenney, 1885). In 1934 it is mentioned in an article relating the contents of a lecture Edward Rowe Snow gave to the Bostonian Society: "Lore of Harbor Islands Abounds with Romance," *Boston Globe*, November 21, 1934. It also appears in his early books *The Islands of Boston Harbor: Their History and Romance, 1626–1935* (Andover, MA: Andover Press, 1935) and *Historic Fort Warren*.
7. Shurtleff, *Topographical and Historical Description of Boston*, 556.
8. "3 Harbor Islands Sold but MDC May Get Two of Them," *Boston Globe*, November 23, 1957.
9. John R. Gillis, *The Human Shore: Seacoasts in History* (Chicago: University of Chicago Press, 2012), 131.
10. Henry David Thoreau, *Cape Cod* (Boston: James R. Osgood, 1877), 2; John T. Cumbler, "Building a Tourist Landscape in a Fragile Ecosystem: Cape Cod in the Nineteenth and Twentieth Centuries," in *A Landscape History of New England*, ed. Blake A. Harrison and Richard W. Judd (Cambridge, MA: MIT Press, 2011), 216–19.
11. "Boston Harbor's Outer Isles," *Boston Daily Globe*, September 9, 1900.
12. Sweetser, *King's Handbook of Boston Harbor*, 180.
13. "Boston Harbor's Outer Isles."
14. Stark, *Illustrated History of Boston Harbor*, 110–13.
15. Stephanie Schorow, "Roughing It on Great Brewster: Daring 19th Century Women Spend Fortnight on 'Enchanted Isle,'" *Harvard Gazette*, April 16, 2009, https://news.harvard.edu.
16. "Boston Harbor's Outer Isles"; Stark, *Illustrated History of Boston Harbor*, 136; *The New England Historical and Genealogical Register*, vol. 48 (Boston: New England Historical and Genealogical Society, 1894), 228–29.
17. Barry Cadigan, "Their Very Own Island: But Trash Barge Spoils the Fun," *Boston Globe*, June 11, 1967.
18. Fred Pillsbury, "Owners All Favor Recreation: Private Islands Would Adapt to Any Plan," *Boston Globe*, November 24, 1968.
19. The temporary Park Commission, founded in 1892, was succeeded by a permanent Metropolitan Park Commission in 1893.
20. Charles William Eliot, *Charles Eliot, Landscape Architect* (1902; reprint, Amherst:

University of Massachusetts Press, 1999), 354; Michael Rawson, *Eden on the Charles: The Making of Boston* (Cambridge, MA: Harvard University Press, 2014), 234.

21. Eliot, *Charles Eliot, Landscape Architect*, 385, 393.
22. Eliot, *Charles Eliot, Landscape Architect*, 595, 529–30.
23. Rawson, *Eden on the Charles*, 265.
24. Eliot, *Charles Eliot, Landscape Architect*, 410.
25. Frederick L. Olmsted, "Report of the Landscape Architect Advisory to the Park Commissioners, 30 December 1887," in *Thirteenth Annual Report of the Board of Commissioners of the Department of Parks for the City of Boston* (Boston: City of Boston, 1887), 57.
26. Olmsted, "Landscape Architect Advisory to the Park Commissioners," 59.
27. Thomas H. O'Connor, *Building a New Boston: Politics and Urban Renewal, 1950–1970* (Boston: Northeastern University Press, 1993), 70.
28. "Shall Trees Be Planted on Boston Harbor Islands?," *Boston Daily Globe*, November 7, 1909.
29. "Asks C.C.C. Work on Harbor Islands," *Boston Globe*, September 9, 1933; "C.C.C. to Reforest Islands in Harbor," *Boston Globe*, November 5, 1933; "To Forest Islands in Boston Harbor," *Boston Globe*, March 26, 1934.
30. "Asks C.C.C. Work on Harbor Islands."
31. "Would Floral Islands Offset Harbor Odors?," *Boston Globe*, May 11, 1934.
32. "Oppose Planting of Exotics in Region," *Boston Globe*, January 24, 1935.
33. "Harbor Police Find 153 Live on Islands," *Boston Globe*, January 3, 1937.
34. "Harbor Dwellers Receive Visit: Bundles of Food Delivered to Islands, Lighthouses," *Boston Herald*, January 2, 1937; Sara H. Hunter, "Edward Rowe Snow: Brief Life of a 'Flying Santa,' 1902–1982," *Harvard Magazine* (January–February 2012), https://harvardmagazine.com.
35. Nicole Critchley, "Edward Rowe Snow," *Boston Athenaeum* (January 2017), https://www.bostonathenaeum.org; Hunter, "Edward Rowe Snow"; Edward Rowe Snow, *The Islands of Boston Harbor*, with the assistance of Jeremy D'Entremont (Carlisle, MA: Applewood Books, 2002), xv–xvi.
36. Snow, *Islands of Boston Harbor* (2002), xv.
37. Godfrey Baldacchino, "The Lure of the Island: A Spatial Analysis of Power Relations," *Journal of Marine and Island Cultures* 1 (2012): 55; Steven Mentz, "Toward a Blue Cultural Studies: The Sea, Maritime Culture, and Early Modern English Literature," *Literature Compass* 6, no. 5 (2009): 1001. For a discussion of New England landscape and the romantic imagination, see Richard W. Judd, *Second Nature: An Environmental History of New England* (Amherst: University of Massachusetts Press, 2014), 143–74.
38. Snow, *Secrets of the North Atlantic Islands*, v. For a discussion of islands in literature, see, for instance, Arndt Kremer, "Eine Welt für sich: Die Insel als literarischer und sprachlicher Grenz- und Denkraum," *Aus Politik und Zeitgeschichte* 68, nos. 32–33 (2018): 47–52; or Diana Loxley, *Problematic Shores: The Literature of Islands* (London: Macmillan, 1990).
39. Edward Rowe Snow, *Mysteries and Adventures along the Atlantic Coast* (New York: Dodd, Mead, 1948), 136–37.
40. Snow, *Secrets of the North Atlantic Islands*.
41. Edward Rowe Snow, *Marine Mysteries and Dramatic Disasters of New England* (New York: Dodd, Mead, 1976); *Ghosts, Gales and Gold* (New York: Dodd, Mead, 1972); and *Piracy, Mutiny, and Murder* (New York: Dodd, Mead, 1959).
42. Snow, *Secrets of the North Atlantic Islands*, v.

43. John R. Gillis, *Islands of the Mind: How the Human Imagination Created the Atlantic World* (New York: Palgrave Macmillan, 2004), 129–30.
44. Snow, *Secrets of the North Atlantic Islands*, v, 58; Snow, *Islands of Boston Harbor, 1630–1971*, 79.
45. All of these stories are found—in slightly different retellings—in *The Romance of Boston Bay*, *Secrets of the North Atlantic Islands*, and *Islands of Boston Harbor, 1626–1935*.
46. Snow, *Secrets of the North Atlantic Islands*, 68. The story was later also related in the memoirs of Charles Jennings's son Harold. See Harold B. Jennings, *A Lighthouse Family: Growing up at the Lovells Island Range Lights*, 2nd ed. (Orleans, MA: Lower Cape, 2015), 65.
47. Snow, *Islands of Boston Harbor, 1630–1971*, 114–19.
48. Schorow, *East of Boston*, 38; Hunter, "Edward Rowe Snow."
49. "3 Harbor Islands Sold but MDC May Get Two of Them"; "Old Boston Forts Go under Hammer," *New York Times*, November 23, 1957.
50. Schorow, *East of Boston*, 39.
51. "Lore of Harbor Islands Abounds with Romance."
52. The three editions appeared in 1935, 1971, and 2002, under slightly varying titles and with some additions and updates. See Snow, *Islands of Boston Harbor, 1626–1935*; *Islands of Boston Harbor, 1630–1971*; and *Islands of Boston Harbor* (2002).
53. Patrick J. Connelly, *Islands of Boston Harbor, 1630–1932: Green Isles of Romance* (Dorchester, MA: Chapple, 1932), 9.
54. Gall Marsh, interview by Boston Harbor Islands Mass. Memories Road Show.
55. Marilyn Turk, "Meet Lighthouse Person Jeremy D'Entremont," *Pathways of the Heart*, January 23, 2015, http://pathwayheart.com.
56. Francis W. Sargent, Review of *The Islands of Boston Harbor*, 1971, box 8-48: Islands of Boston Harbor, Edward Rowe Snow Collection, Howard Gotlieb Archival Research Center at Boston University Libraries.

Chapter 5: A New Spectacle

1. Suzanne Gall Marsh, interview by the Boston Harbor Islands Mass. Memories Road Show, September 17, 2011, University Archives and Special Collections, Joseph P. Healey Library, University of Massachusetts, Boston, https://vimeo.com/103919924; *Friends of the Boston Harbor Islands: Forty Years of Public Service on Land and Sea* (Boston: Friends of the Boston Harbor Islands, 2019), 7; Ruth Aronson, "An Island, Waiting for You," *Boston Globe*, September 11, 1977.
2. Stephanie Schorow, *East of Boston: Notes from the Harbor Islands* (Charleston, SC: History Press, 2008), 97–101; "Gallops Island," Boston Harbor Islands National Recreation Area, National Park Service, https://www.nps.gov/boha/learn/historyculture/facts-gall.htm; "3 Harbor Islands to Be Fixed Up this Summer," *Boston Globe*, March 24, 1975.
3. *Friends of the Boston Harbor Islands*, 7; Katherine F. Abbott, interview by the author, January 28, 2021; Mari Megias, "Full Circle: Kathy Abbott MC/MPA 1988 Is Creating Change for Boston Harbor," Harvard Kennedy School, January 19, 2017, https://www.hks.harvard.edu/more/alumni/alumni-stories/full-circle.
4. *Friends of the Boston Harbor Islands*, 15.
5. For the history of U.S. environmental legislation before the 1970s, see Karl Boyd Brooks, *Before Earth Day: The Origins of American Environmental Law, 1945–1970* (Lawrence: University Press of Kansas, 2009).
6. The literature about the 1970s North American environmental movement is large and

growing. David Stradling, *The Environmental Moment, 1968–1972* (Seattle: University of Washington Press, 2012), offers an overview of the most important contemporary texts. Joachim Radkau, *The Age of Ecology: A Global History* (Cambridge: Polity, 2014), provides a global context of the environmental movement, while Robert Gottlieb, *Forcing the Spring: The Transformation of the American Environmental Movement*, rev. ed. (Washington, DC: Island Press, 2005), links the 1970s environmental movement to its predecessors; for the 1960s roots of 1970s environmentalism, see Adam Rome, "'Give Earth a Chance': The Environmental Movement and the Sixties," *Journal of American History* 90, no. 2 (2003). Other notable publications on the topic include Kirkpatrick Sale, *The Green Revolution: The American Environmental Movement, 1962–1992* (New York: Hill and Wang, 1993); and Hal K. Rothman, *The Greening of a Nation? Environmentalism in the United States since 1945* (Fort Worth: Harcourt Brace College, 1998).

7. Constitution of the Commonwealth of Massachusetts, Article XCVII, https://malegislature.gov/Laws/Constitution#amendmentArticleXCVII.
8. For the historical and regional context of Massachusetts environmental legislation, see Richard W. Judd, *Second Nature: An Environmental History of New England* (Amherst: University of Massachusetts Press, 2014), 240–71.
9. Boston Harbor Islands Commission Conference, 1968–1969, Conference Transcript—Workshop 4: Ecology/Oceanology, June 24, 1969, folder 12, box 2, series II, David C. Harrison Papers, State Library of Massachusetts Special Collections.
10. *Third and Final Report of the Special Commission on the Boston Harbor Islands: Massachusetts House of Representatives Resolve No. 5051* (Boston: Wright and Potter, 1970), 60, 38, 7.
11. Scott Carlin, "Rediscovering Boston Harbor: The Political Geography of Urban Environmental Agendas" (PhD diss., Clark University, 1995), 177–88; *Friends of the Boston Harbor Islands*, 13.
12. Save Our Shores, Dorchester Bay Boston Harbor Islands Project, folder 31: Boston Harbor Islands Correspondence, February–April 1969, box 1, Mayor Kevin H. White Records, City of Boston Archives.
13. Gall Marsh, interview by Boston Harbor Islands Mass. Memories Road Show.
14. Aronson, "An Island, Waiting for You."
15. Anne Raver, "An Island of Calm," *Boston Herald*, May 12, 1983; Laura Bartlett, "So Much for the Harbor Islands," *Offshore Magazine* (August 1984). The story of the crew of a French ship having been killed by local Natives at Peddocks Island in the 1610s first appears in Thomas Morton's 1637 book *The New English Canaan*. It is probably fabricated, but has been picked up by M. F. Sweetser in his *King's Handbook of Boston Harbor*, 3rd ed. (Boston: Moses King, 1889) and by later authors such as Edward Rowe Snow or David Kales. Snow, *The Islands of Boston Harbor, 1630–1971*, 2nd ed. (New York: Dodd, Mead, 1971); Kales, *Boston Harbor Islands: A History of an Urban Wilderness* (Charleston, SC: History Press, 2007), 21.
16. Kevin H. White, Boston in the 1980s, folder 23: Development: Boston in the 1980s, box 124, White Records.
17. *Third and Final Report of the Special Commission on the Boston Harbor Islands*, 38.
18. "The Politics of Sewage," *Patriot Ledger*, May 18, 1985.
19. "Legislators Push Deer Island Plant," *Patriot Ledger*, May 23, 1985.
20. "Protecting Long Island," *Boston Globe*, October 15, 1983.
21. For a detailed discussion of the term, see James W. Feldman, *A Storied Wilderness: Rewilding the Apostle Islands* (Seattle: University of Washington Press, 2011).
22. *Third and Final Report of the Special Commission on the Boston Harbor Islands*, 10.

23. Sweetser, *King's Handbook of Boston Harbor*, 249; Mark A. Berhow, *American Seacoast Defenses: A Reference Guide*, 3rd ed. (McLean, VA: Coast Defense Study Group Press, 2015); Julie A. Richburg and William A. Patterson, "Historical Description of the Vegetation of the Boston Harbor Islands, 1600–2000," *Northeastern Naturalist* 12, no. 3 (2005); Barry Cadigan, "Their Very Own Island: But Trash Barge Spoils the Fun," *Boston Globe*, June 11, 1967.
24. James B. Ayres, "Boston Harbor: After Five Years of Controversy, Another Master Plan Is Being Made," *Boston Globe*, February 28, 1971.
25. Kenneth D. Campbell, "Legislature Takes Action on Environment, Housing," *Boston Globe*, August 23, 1970.
26. *Third and Final Report of the Special Commission on the Boston Harbor Islands*, 22, 46.
27. Boston Harbor Islands Advisory Commission, Advisory Commission Meeting, Discussion Summary, July 1976, box 32: Boston Harbor Islands Advisory Committee, 1975–76, Boston 200 Records, City of Boston Archives; *Boston Harbor Islands: Report of a Special Resource Study* (Boston: National Park Service, North Atlantic Region, 1994), 5.
28. Michael Kenney, "State Acquires 9 Harbor Islands for Public Use," *Boston Globe*, April 21, 1973.
29. Boston Harbor Islands Week, July 1976, box 32: Boston Harbor Islands Advisory Committee, 1975–76, Boston 200 Records.
30. Metropolitan Area Planning Council, *Boston Harbor Islands Comprehensive Plan*, October 1972 (Boston: MAPC), 11, 31, 53–54.
31. Metropolitan Area Planning Council, *Boston Harbor Islands Comprehensive Plan*, 54–55.
32. "The Case for Super Harbor 1," *Boston Globe*, May 27, 1984.
33. Boston Harbor Islands State Park: Draft Master Plan Update, May 31, 1984, folder 38: Boston Harbor Islands: Draft Concept Plan, box 1, series I: Administrative Materials, 1980–88, Special Legislative Commission on the Development of Boston Harbor Records, State Library of Massachusetts Special Collections.
34. Massachusetts Department of Environmental Management, *Boston Harbor Islands State Park 1986 Master Plan* (Boston: DEM, 1986), 3.
35. Katherine Abbott, DEM Office of Planning and Development, to Secretary James Hoyte and Jim Gutensohn, June 1983, folder Boston Harbor Islands, box 1: 1980–86 Boston Harbor A–L, EN 1/Series 578 Massachusetts, Issue and Policy Subject Files, 1980–93, Executive Office of Environmental Affairs, Massachusetts Archives.
36. Boston Harbor Islands State Park: Draft Master Plan Update, May 31, 1984.
37. The problem was addressed by a series of reports and special committees. For instance, see Joint Board Acting under the Provisions of Chapter 56 of the Resolves of 1917, *Report on an Investigation of the Discharge of Sewage into Boston Harbor* (Boston: Wright and Potter, 1918), House No. 1215; Special Commission Relative to the Discharge of Sewage into Boston Harbor, *Report under Chapter 29* (Boston: Wright and Potter, 1930), Senate No. 56; and Special Commission on the Discharge of Sewage into Boston Harbor and Its Tributaries, *Report, December 1936* (Boston: Wright and Potter, 1937), House No. 1600.
38. Special Commission Relative to the Discharge of Sewage into Boston Harbor, *Report under Chapter 29*, 55.
39. Special Commission on the Discharge of Sewage into Boston Harbor and Its Tributaries, *Report, December 1936*, 26.
40. Steven M. Rudnick, "Remaking Boston Harbor: Cleaning Up after Ourselves," in *Remaking Boston: An Environmental History of the City and Its Surroundings*, ed. Anthony N. Penna and Conrad E. Wright (Pittsburgh: University of Pittsburgh Press, 2009), 68–69.

41. "Second Interim Report of the Special Commission on the Boston Harbor Islands: Massachusetts House of Representatives Resolve No. 4884" (Boston, 1970), 15–16.
42. Boston Harbor Islands Commission Conference, 1968–69, Conference Transcript—Workshop 1: Pollution, June 24, 1969, folder 9, box 1, series II, Harrison Papers, State Library of Massachusetts Special Collections.
43. David Standley, Boston's Environment, June 1975, folder 16: Environment: Boston 1975, box 127, White Records.
44. "Boston Harbor Not the Only Murky Water," *MAPC Regional Report: The Newsletter of the Metropolitan Area Planning Council, Boston, Massachusetts* (September 1977), 1.
45. The efforts of the Nut Island staff to accomplish their task with inadequate funds, outdated equipment, and without assistance from the MDC eventually led to catastrophic results and became the namesake of a destructive organizational dynamic called the Nut Island effect by Paul Levy, the former head of the Metropolitan Water Resources Authority that took over from the MDC in 1985 as the plant's supervising body. See Paul F. Levy, "The Nut Island Effect: When Good Teams Go Wrong," *Harvard Business Review* (March 2001).
46. Alexander Reid, "A Lesson in Marine Life—and Reality," *Boston Globe*, May 9, 1981; "Harbor Sewage Flow to Pollute Beaches," *Patriot Ledger*, June 7, 1986; Dianne Dumanoski, "Raw or Treated, the Sewage Flows into Boston Harbor," *Boston Sunday Globe*, December 19, 1982.
47. Technical Advisory Group for Boston Harbor and Massachusetts Bay, Massachusetts Executive Office of Environmental Affairs, Study Plan for Basinwide Management of the Boston Harbor/Massachusetts Bay Ecosystem, July 1986, folder 37: Basinwide Management Study Files, box 1, series I: Administrative Materials, 1980–88, Special Legislative Commission on the Development of Boston Harbor Records, State Library of Massachusetts Special Collections.
48. Rudnick, "Remaking Boston Harbor," 70.
49. Charles M. Haar, *Mastering Boston Harbor: Courts, Dolphins, and Imperiled Waters* (Cambridge, MA: Harvard University Press, 2005), 20, 26.
50. Andrew T. Savage, "Boston Harbor: The Anatomy of a Court-Run Cleanup," *Boston College Environmental Affairs Law Review* 22, no. 2 (1995): 372.
51. Dianne Dumanoski, "Court Set to Enforce Harbor Cleanup," *Boston Globe*, September 6, 1985.
52. "Sargent Takes the Harbor On," *Boston Globe*, June 2, 1983; Diane E. Lewis, "New Group to Push for Harbor Cleanup," *Boston Globe*, January 17, 1985.
53. "$4B Harbor Cleanup to Begin," *Boston Herald*, August 8, 1988.
54. Scott Allen, "Closing in on a Healthy Harbor," *Boston Globe*, March 15, 2000.
55. During the 1988 presidential campaign, George H. W. Bush famously derided Boston Harbor as the "harbor of shame," blaming the harbor's state on his rival, Massachusetts governor Michael Dukakis. See Robin Toner, "Bush, in Enemy Waters, Says Rival Hindered Cleanup of Boston Harbor," *New York Times*, September 2, 1988.
56. U.S. Environmental Protection Agency Region 1, *Final Environmental Impact Statement: Siting of Wastewater Treatment Facilities for Boston Harbor*, vol. 1, *Comprehensive Summary* (Boston: EPA, 1985), 2; Johanna Seltz, "Harbor's Long Island Awaits New, Undecided Roles," *Patriot Ledger*, April 9, 1985.
57. Snow, *Islands of Boston Harbor, 1630–1971*, 114, 119.
58. Alan Mikal, *Exploring Boston Harbor* (North Quincy, MA: Christopher, 1973), 84.
59. Metropolitan Area Planning Council, Spectacle Island Support Documentation, March 1973, University Archives and Special Collections, Joseph P. Healey Library, University of Massachusetts, Boston.

60. Draft Boston Harbor Islands State Park Master Plan, 1984, folder Boston Harbor Islands, box 1: 1980–86 Boston Harbor A–L, EN 1/Series 578 Massachusetts, Issue and Policy Subject Files, 1980–93, Executive Office of Environmental Affairs, Massachusetts Archives; Massachusetts Department of Environmental Management, *Boston Harbor Islands State Park 1986 Master Plan*, 21.
61. Discussion Paper: Sludge Management Alternatives for the Metropolitan District Commission, prepared by Cheryl A. Breen, Water Quality Planner, Massachusetts Office of Coastal Zone Management, 1983, folder Boston Harbor Water Quality Committee 2, box 1: 1980–86 Boston Harbor A–L, EN 1/Series 578 Massachusetts, Issue and Policy Subject Files, 1980–93, Executive Office of Environmental Affairs, Massachusetts Archives.
62. Massachusetts Water Resources Authority, News Release: Massachusetts Water Resources Authority Nears Final Decision on Sludge Siting, September 23, 1988, folder 91: Press Releases, box 2, series II: Department of Environmental Quality Engineering, 1990, Special Legislative Commission on the Development of Boston Harbor Records, State Library of Massachusetts Special Collections.
63. Nick Tate, "Towns Say 'No Way!' to Sludge Plan: Officials Vow Fight against Plant Sites," *Boston Herald*, July 29, 1987; Beverly Ford, "Walpole Residents Promise Bitter Fight," *Boston Herald*, July 30, 1987; Doug Boyle, "Deer Island on Final 'Sludge List,'" *Winthrop Sun-Transcript*, July 29, 1987.
64. John Impemba, "Water Authority Rips Towns for Sludge Plan Resistance," *Boston Herald*, July 30, 1987.
65. Richard Pinkofsky, "Operation Oasis Kicks into High Gear," *Brockton News Tribune*, July 13, 1988.
66. Stephen Walsh, "Inland Sludge Site Foes Tour Harbor Island," *Patriot Ledger*, September 2, 1988.
67. Michael T. Cheney, Quincy city councilor, to James Hoyte, MWRA chairman, August 6, 1987, folder 17: Environment, 1988, box I-27, Mayor Raymond L. Flynn Records, City of Boston Archives.
68. Steven J. Fradkin, "Spectacle Island Could Never Be Parkland," *Boston Herald*, August 20, 1988.
69. Executive Office of Environmental Affairs, Publicity Release: Spectacle Island Design Competition, Boston Harbor Islands, 1983, folder Boston Harbor Islands, box 1: 1980–86 Boston Harbor A–L, EN 1/Series 578 Massachusetts, Issue and Policy Subject Files, 1980–93, Executive Office of Environmental Affairs, Massachusetts Archives.
70. Massachusetts Department of Environmental Management, *Boston Harbor Islands State Park 1986 Master Plan*, 20.
71. Massachusetts Water Resources Authority, *Residuals Management Facilities Plan: Draft Report on Environmental Screening Analysis of Candidate Sites*, vol. 2, *Coastal and Island Sites* (Boston: Black & Veatch, 1988), 4.1–4.2; Boston Transportation Department, Summarization of November 18, 1987, Meeting of the Inter-Agency Committee for the Central Artery/Third Harbor Crossing Project, November 30, 1987, folder BRA, Central Artery, Transportation Department, Minutes and News, box III C-7, Flynn Records.
72. Sierra Club New England Chapter, Greater Boston Group, Position on Spectacle Island, October 4, 1988, box 3: Central Artery, EN1/Series 577 Massachusetts, Intergovernmental Subject Files, Executive Office of Environmental Affairs, Massachusetts Archives.
73. Friends of the Boston Harbor Islands, Comments re: MDPW proposal for fill on

Spectacle Island, undated [probably August 6, 1990], Advocacy—Spectacle Island, CA/T project, Suzanne Gall Marsh and George Marsh private collection.

74. Larry Tye, "$6.1B Cleanup of Boston Harbor Gets under Way: Deer Island Groundbreaking First Step in 11-Year Project," *Boston Globe*, August 11, 1988.
75. Eve Epstein, "Bulger Issues Spectacle Island Warning at Ground-Breaking for Sewer Plant," *Patriot Ledger*, August 11, 1988; Norvin Leach, "Bulger Vows Battle on Island Sludge Site," *Malden Evening News*, August 11, 1988; "Bulger Opposed to Spectacle Island Sludge Site," *Woburn Daily Times and Chronicle*, August 11, 1988; Thor Jourgensen, "Bulger Says 'No' to Sludge at Island," *Lynn Daily Evening Item*, August 11, 1988; Sue Reinert, "Flynn, Bulger Lock Horns over Island," *Boston Business Journal*, August 15, 1988.
76. Sarah Just, "Official Suggests Town Not Upset Bulger," *Patriot Ledger*, July 27, 1988.
77. Mark S. Morrow, "Boston Studies Harbor Island as Sludge Site," *Patriot Ledger*, May 6, 1988.
78. Rudnick, "Remaking Boston Harbor," 72.
79. Special Commission on the Development of Boston Harbor, Minutes, November 21, 1985, folder 30: Commission Meetings 1985, box 1, series I: Administrative Materials, 1980–88, Special Legislative Commission on the Development of Boston Harbor Records, State Library of Massachusetts Special Collections.
80. Boston Harbor Supplemental Draft EIS: Wastewater Facility Siting Alternatives, folder 26: Commission Meetings 1983 II, box 1, series I: Administrative Materials, 1980–88, Special Legislative Commission on the Development of Boston Harbor Records, State Library of Massachusetts Special Collections; U.S. Environmental Protection Agency Region 1, *Final Environmental Impact Statement*.
81. Secretary John DeVillars to Governor Michael Dukakis, March 1990, Peddocks Island, box 3: MDC [Peddocks Island], EN1/Series 577 Massachusetts, Intergovernmental Subject Files, Executive Office of Environmental Affairs, Massachusetts Archives.
82. U.S. Environmental Protection Agency Region 1, *Siting of Wastewater Treatment Facilities: Summary of Supplemental Draft Environmental Impact Statement* (Boston: EPA, 1984).
83. Haar, *Mastering Boston Harbor*, 298.
84. Draft Status of Dredged and Excavated Material Disposal Options for Boston Harbor, 1988, folder 35: Commission Meetings 1988, box 1, series I: Administrative Materials, 1980–88, Special Legislative Commission on the Development of Boston Harbor Records, State Library of Massachusetts Special Collections.
85. *Report of a Special Resource Study*; Omnibus Parks and Public Lands Management Act of 1996, Public Law 104-333, Sec. 1029: Boston Harbor Islands National Recreation Area, November 12, 1996; Abbott, interview by the author.
86. *Report of a Special Resource Study*, 12, 23.
87. *Report of a Special Resource Study*, 23; New England Aquarium, Spectacle Island: Statement of Interest, June 25, 1998, SIPAC files, Gall Marsh and Marsh private collection; Omnibus Parks and Public Lands Management Act of 1996.
88. John Ballard Blake, *Public Health in the Town of Boston, 1630–1822* (Cambridge, MA: Harvard University Press, 1959), 18.
89. Christopher Klein, "A Harbor Spectacle," *Boston Globe*, June 22, 2008.
90. Special Commission on the Development of Boston Harbor, minutes, December 3, 1987, folder 3: Activities, series I: Administrative Materials, 1980–88, Special Legislative Commission on the Development of Boston Harbor Records, State Library of Massachusetts Special Collections.

91. Anthony Flint, "10 Years Later, Did the Big Dig Deliver?," *Boston Globe*, January 3, 2016; Dan McNichol, *The Big Dig* (New York: Silver Lining Books, 2001), 87.
92. David D'Alessandro, "The Bright Side of the Big Dig," *Boston Globe*, September 24, 2007; "The Big Dig: Facts and Figures," https://www.mass.gov/info-details/the-big-dig-facts-and-figures.
93. "The Big Dig: Facts and Figures."
94. Frederick P. Salvucci, Central Artery/Third Harbor Tunnel Update, May 20, 1988; Boston Transportation Department, Summarization of November 18, 1987 Meeting of the Inter-Agency Committee for the Central Artery/Third Harbor Crossing Project, November 30, 1987; both folder BRA, Central Artery, Transportation Department, Minutes and News, box III C-7, Flynn Records.
95. Boston Transportation Department, Comments on the Draft Supplemental Environmental Impact Statement for the CA/T Project, August 22, 1990, folder Transportation Department, CA/T: Comments on Environmental Impact Statement, August–September 1990, box IV C-22, Flynn Records.
96. Lorraine M. Downey, Central Artery/Third Harbor Tunnel Project, August 8, 1990, folder Transportation Department, CA/T: Comments on Environmental Impact Statement, August–September 1990, box IV C-22, Flynn Records.
97. Massachusetts Department of Environmental Management, *Boston Harbor Islands State Park 1986 Master Plan*, 20.
98. Special Commission on the Development of Boston Harbor, Minutes, April 30, 1987 Meeting, folder 33: Commission Meetings 1987 II, box 1, series I: Administrative Materials, 1980–1988, Special Legislative Commission on the Development of Boston Harbor Records, State Library of Massachusetts Special Collections.
99. Morrow, "Boston Studies Harbor Island as Sludge Site."
100. Peter J. Howe, "Artery Project Engineers Dig for Answers on Dirt Dumping," *Boston Globe*, March 13, 1988.
101. James Gutensohn, Announcement of Boston Harbor Islands Master Plan, Formation of Enterprise, November 30, 1984. Summary of Major Points for Governor's Remarks, folder Boston Harbor Islands, box 1: 1980–1986 Boston Harbor A–L, EN 1/Series 578 Massachusetts, Issue and Policy Subject Files, 1980–1993, Executive Office of Environmental Affairs, Massachusetts Archives.
102. "Deadly Dump," *Access: Central Artery/Third Harbor Tunnel Project* (Winter 1988): 2.
103. Artery Business Committee, Evaluation of the Central Artery/Tunnel Project: Supplemental Environmental Impact Statement, June 1990, folder Transportation Department, CA/T: Comments on Environmental Impact Statement, August–September 1990, box IV C-22, Flynn Records.
104. Central Artery SDEIS Meeting, March 19, 1990, box 3: Central Artery, EN1/Series 577 Massachusetts, Intergovernmental Subject Files, Executive Office of Environmental Affairs, Massachusetts Archives.
105. McNichol, *The Big Dig*, 92.
106. Boston Redevelopment Authority, Comments from the BRA on the Final Supplemental Environmental Impact Statement/Report, December 24, 1990, box 3: Central Artery. EN1/Series 577 Massachusetts, Intergovernmental Subject Files, Executive Office of Environmental Affairs, Massachusetts Archives; Massachusetts Department of Public Works, *Final Supplemental Environmental Impact Report: Central Artery—Boston, Massachusetts*, pt. 1, bk. 1 (Boston: DPW, 1990), 16; Massachusetts Department of Public Works, *Final Supplemental Environmental Impact Report: Central Artery—Boston, Massachusetts*, pt. 2 (Boston: DPW, 1990), IIB 4.1.
107. Environmental Notification Form: Gardner Street Landfill Closure and Reuse

Project, March 1997, box 1: Landfill Files, Office of Special Projects (Alice Hennessey Records), Mayor Thomas M. Menino Records, City of Boston Archives.

108. John Russell, "Boston Considers Park on Landfill," *Waste News*, December 11, 1995.
109. Environmental Notification Form: Gardner Street Landfill Closure and Reuse Project, March 1997.
110. Susan O'Neill, "From Trash Heap to Treasure," *West Roxbury Transcript*, November 17, 1999.
111. Thomas Grillo, "Park Construction Steaming Along," *Boston Globe*, November 14, 1999; Scott Allen, "Downtown Islands: Boston Harbor's Much-Abused Islands May Be Getting a Face-Lift," *Island Journal* (1995): 22.
112. Scott Allen, "Creating a Spectacle: Dumping Ground Rising as a Harbor Park Showplace," *Boston Globe*, May 31, 1997.
113. Andrew Blake, "Debris Soon to Be Spectacle's Latest Layer," *Boston Globe*, September 22, 1992.
114. Thomas C. Palmer, "Getting the Dirt on the Big Dig: A New Profile Shapes Up of Harbor Island," *Boston Globe*, October 23, 1995.
115. Thomas C. Palmer, "Harbor Island Gets New View," *Boston Globe*, November 27, 1994; Palmer, "Getting the Dirt on the Big Dig."
116. "A Bridge Too Snarled," *Boston Globe*, September 16, 1989; Beth Daley, "A Spectacular Rebirth: Big Dig Dirt Has Transformed Harbor Island into Parkland," *Boston Globe*, August 24, 2003.
117. Allen, "Creating a Spectacle."
118. " Boston Harbor Map," *Boston Harbor Islands State Park Newsletter: Massachusetts Department of Environmental Management* (September 1983): 5.
119. Raphael Lewis, "Eyesore No More: Eight Years and 7 Million Tons of Big Dig Dirt Later, Island Once Used as Dump Almost Ready for Visitors," *Boston Globe*, February 17, 2002.
120. Klein, "A Harbor Spectacle."
121. Ari Bloomekatz, "Spectacle Island Completes Its Latest Transformation: Former City Dump Reopens as a Park," *Boston Globe*, June 25, 2006.

Epilogue

1. "Things to Do 2016," *Harbor Islands Times* (2016 Season): 2.
2. Boston Harbor Islands Partnership, *Boston Harbor Islands, a National Park Area: General Management Plan* (Boston: National Park Service, 2002), 55; "Things to Do 2016," 2, 12.
3. The park maintains two principal websites, one run by the National Park Service and the other by the Boston Harbor Islands Partnership. Both give information about island geography, histories, and park news and activities. See "Boston Harbor Islands National Recreation Area (U.S. National Park Service)," https://www.nps.gov/boha/index.htm; and "Boston Harbor Islands National & State Park," http://www.bostonharborislands.org/.
4. "Save the Harbor," http://www.savetheharbor.org/; "Friends of the Boston Harbor Islands Established 1979," http://www.fbhi.org/index.html; Kathy Abbott, video interview, *Charles River Conservancy Parkland Show*, September 13, 2017, https://www.youtube.com/watch?v=SLk9zTKtXBY.
5. William P. Coughlin, "Beware of Touching Hub's Outer Islands," *Boston Globe*, January 24, 1971; Save the Brewsters, "Save Outer Brewster: Frequently Asked Questions," October 24, 2005, LNG Proposal Outer Brewster 2005, box 3, Suzanne

Gall Marsh Records, University Archives and Special Collections, Joseph P. Healey Library, University of Massachusetts, Boston.

6. Richard Pinkofsky, "Operation Oasis Kicks into High Gear," *Brockton News Tribune*, July 13, 1988.
7. Christopher Klein, *Discovering the Boston Harbor Islands: A Guide to the City's Hidden Shores*, 2nd ed. (Boston: Union Park Press, 2011), 15–16.
8. Boston Harbor Islands Partnership, *Boston Harbor Islands*, 47–48.
9. "2016: A Year to Celebrate," *Harbor Islands Times* (2016 Season): 1.
10. Mara Vorhees, *Boston*, 4th ed. (London: Lonely Planet, 2009), 9; Barbara Rockwell, *Boston*, 5th ed. (Singapore: APA, 2013), 132.
11. "Constitution Marina's Island Moorings: Escape to the Boston Harbor Islands" (undated).
12. "Core Sample—2007—Teri Rueb," http://terirueb.net/core-sample-2007/.
13. Martin V. Melosi, "Fresh Kills: The Making and Unmaking of a Wastescape," in *Out of Sight, Out of Mind: The Politics and Culture of Waste*, ed. Christof Mauch, RCC Perspectives: Transformations in Environment and Society (Munich: Rachel Carson Center, 2016), 60; Melosi, *Fresh Kills: A History of Consuming and Discarding in New York City* (New York: Columbia University Press, 2020), 3.
14. Eric J. Fauss, "The Wild Lands of Gotham: City and Nature in Jamaica Bay, New York, 1884–1990" (PhD diss., University of Connecticut, 2014).
15. "Hart-Miller Island State Park," https://dnr.maryland.gov/publiclands/Pages/central/hartmiller.aspx.
16. Matthew Morse Booker, *Down by the Bay: San Francisco's History between the Tides* (Berkeley: University of California Press, 2013).
17. Stéphane Castonguay and Michèle Dagenais, eds., *Metropolitan Natures: Environmental Histories of Montreal* (Pittsburgh: University of Pittsburgh Press, 2011).
18. Derek Lamb and Lawrence Rosenblum, *Boston: Harbor/City/Islands*, produced by Stanley Jacks, available courtesy of University Archives and Special Collections, Joseph P. Healey Library, University of Massachusetts, Boston, SC-0175: Volunteers and Friends of the Boston Harbor Islands Inc. records (Cambridge, MA: Massachusetts Institute of Technology, 1969).

INDEX

PAVLA ŠIMKOVÁ is postdoctoral researcher at the Rachel Carson Center for Environment and Society at Ludwig-Maximilians-Universität (LMU) in Munich, Germany. She studied American cultural history, English literature, and political science at the Masaryk University in Brno, Czech Republic, and at LMU Munich where she received her doctorate in 2019. Šimková worked as a research associate at the Rachel Carson Center from 2012 until 2018. She has been a coauthor of the *Biographisches Lexikon zur Geschichte der böhmischen Länder* at the Collegium Carolinum in Munich since 2011. She is currently one of the editors of the journal *WerkstattGeschichte.* Her research interests include American and Central European environmental history and urban history.